# ARTIFICIAL NUTRITION AND HYDRATION

# Philosophy and Medicine

## VOLUME 5

*Founding Co-Editor*
Stuart F. Spicker

*Senior Editor*

H. Tristram Engelhardt, Jr., *Department of Philosophy, Rice University, and Baylor College of Medicine, Houston, Texas*

*Associate Editor*

Lisa M. Rasmussen, *Department of Philosophy, University of North Carolina at Charlotte, North Carolina*

## CATHOLIC STUDIES IN BIOETHICS

*Series Founding Co-Editors*

John Collins Harvey, *Georgetown University, Washington, D.C., U.S.A.*
Francesc Abel, *Institut Borja de Bioetica, Center Borja, Barcelona, Spain*

*Series Editor*

Christopher Tollefsen, *University of South Carolina, Columbia, SC, U.S.A.*

*Editorial Advisory Board*

Joseph Boyle, *St. Michael's College, Toronto, Canada*
Thomas Cavanaugh, *University of San Francisco, San Francisco, CA, U.S.A.*
Mark Cherry, *St. Edward's University, Austin, TX, U.S.A.*
Ana Smith Iltis, *St. Louis University, St. Louis, MO, U.S.A.*

# ARTIFICIAL NUTRITION AND HYDRATION

## THE NEW CATHOLIC DEBATE

*Edited by*

CHRISTOPHER TOLLEFSEN

*University of South Carolina, Columbia, SC, USA*

TX
353
A78
2008
Web

Springer

A C.I.P. Catalogue record for this book is available from the Library of Congress.

ISBN 978-1-4020-6206-3 (HB)
ISBN 978-1-4020-6207-0 (e-book)

Published by Springer,
P.O. Box 17, 3300 AA Dordrecht, The Netherlands.

*www.springer.com*

*Printed on acid-free paper*

All Rights Reserved
© 2008 Springer
No part of this work may be reproduced, stored in a retrieval system, or transmitted
in any form or by any means, electronic, mechanical, photocopying, microfilming, recording
or otherwise, without written permission from the Publisher, with the exception
of any material supplied specifically for the purpose of being entered
and executed on a computer system, for exclusive use by the purchaser of the work.

# Preface

Pope John Paul II surprised much of the medical world in 2004 with his strongly worded statement insisting that patients in a persistent vegetative state should be provided nutrition and hydration (John Paul II, 2004). While many Catholic bioethicists defended the Pope's claim that the life of all human beings, even those in a persistent vegetative state or a coma, was worth protecting, others argued that the Pope's position marked a shift from the traditional Catholic teaching on the withdrawal of medical treatment at the end of life.

The debate among Catholic bioethicists over the Pope's statement only grew more intense during the controversy surrounding Terri Schiavo's death in 2005, as bioethicists on both sides of the debate argued about the legitimacy of removing her feeding tubes. Many Catholics were troubled by the Florida courts' reliance on the testimony of Schiavo's husband regarding her wishes, given his apparent neglect of her, and his new relationship with another woman. Moreover, Schiavo's family expressed repeatedly and strongly their willingness to provide care for her. Accordingly, to many, it seemed that the removal of her feeding tubes was an act of euthanasia.

Nevertheless, the Catholic tradition firmly asserts the right of patients to refuse medical treatment when such treatment is "extraordinary" or "disproportionate" (Pius XII, 1957). So, while Schiavo's treatment seemed egregious, and not in accordance with John Paul II's allocution, it still seemed open to some Catholic theologians and philosophers to argue for the legitimacy of removal in her case, and others. The controversy thus continued.

This volume takes stock of that controversy, and the Papal *Allocution* that played a considerable part in its generation. In this volume, philosophers and moral theologians address both the interpretive issue: What, precisely, was the Pope forbidding, and requiring? And they address the moral issue: What, precisely, is owed to patients in a persistent vegetative state? When, if ever, is it permissible to remove their feeding tubes? When is such removal tantamount to euthanasia? Philosophers and theologians on both sides of the issue take stock of the Pope's *Allocution*, the weight of tradition, and the strength of the arguments (see especially Gomez-Lobo, 2008; Boyle, 2008; Garcia, 2008; Cataldo, 2008). *Artificial Nutrition and Hydration: The New Catholic Debate* thus provides a helpful roadmap to one of the most difficult issues of Catholic bioethics today.

The essays in this volume do more than this, however. The essays in this book go beyond the philosophical and theological controversies concerning ANH, in some cases to situate the debate in terms of Catholic and moral understandings of the importance of food for community, or the relationship between a community and its disabled (Fisher, 2008; Degnan, 2008). Other essays provide an account of the history of the debate and the status of the law regarding the feeding of patients in a vegetative state (May, 2008; Laing, 2008). There is also a symposium on the position of Fr. Kevin O'Rourke, whose work on ANH has been extremely influential, but also controversial, over the past two decades (O'Rourke, 2008a; 2008b; Latkovic, 2008; Lee, 2008).

In sum, *Artificial Nutrition and Hydration: The New Catholic Debate* provides a comprehensive introduction to the issue, and illustrates the work of some of the Church's finest philosophical and theological minds at work in resolving the moral issues at stake in this difficult problem.

It is important to note one development in the discussion which has occurred while this book was being put in press, and which has thus not been addressed by any of its contributors. On August 1, 2007, the Congregation for the Doctrine of the Faith released a brief document titled "Responses to Certain Questions of the United States Conference of Catholic Bishops Concerning Artificial Nutrition and Hydration" (CDF, 2007). Approved by Pope Benedict XVI, the document answers two questions. First, as to the question of whether "the administration of food and water (whether by natural or artificial means) to a patient in a 'vegetative state' [is] morally obligatory except when they cannot be assimilated by the patient's body or cannot be administered to the patient without causing significant physical discomfort" (CDF, 2007, q. 1), the document answers affirmatively.

In its answer to this question, the CDF follows John Paul II in affirming that "the administration of food and water even by artificial means is, in principle, an ordinary and proportionate means of preserving life. It is therefore obligatory to the extent to which, and for as long as, it is shown to accomplish its proper finality, which is the hydration and nourishment of the patient" (CDF, 2007, r. 1).

The document then asks, "When nutrition and hydration are being supplied by artificial means to a patient in a 'permanent vegetative state,' may they be discontinued when competent physicians judge with moral certainty that the patient will never recover consciousness" (CDF, 2007, q. 2)? The document denies this: "A patient in a 'permanent vegetative state' is a person with fundamental human dignity and must, therefore, receive ordinary and proportionate care which includes, in principle, the administration of water and food even by artificial means" (CDF, 2007, r.2).

These two responses appear to confirm many claims made by contributors to this volume regarding the proper interpretation of the Papal *Allocution*, and the obligation to provide nutrition and hydration to the permanently unresponsive. At the same time, it is important to note a question that arises about the CDF document: does it consider only the narrow of question of patients who are in a permanent vegetative state and their caregivers as such, without making a judgment about advance directives from patients regarding termination of their nutrition and hydration? The document seems to assert unequivocally a duty to feed as long as it is not futile; one important question is thus whether that duty would override a duty to respect a patient's advance directive in case

# Part I
# The Issue

# Chapter 1
# Why do Unresponsive Patients Still Matter?[1]

Bishop Anthony Fisher, O.P.

## 1.1 Civilization After Schiavo?

### 1.1.1 Introduction to the Contest

She was not brain-dead, not dying, not comatose, not on 'life-support.' A potassium imbalance in 1990 had led to a collapse and a coma from which she emerged severely cognitively impaired, if otherwise quite healthy. She was diagnosed as being in a 'persistent vegetative state' ('PVS') though some believed she responded in a rudimentary way to her family and environment. What no-one contested was that she was being fed and hydrated through a tube. Her estranged husband-guardian wanted that feeding stopped; her parents and other family members wanted it continued. Her husband said he thought she had had enough and clearly had had enough himself; her family thought the pious Catholic woman would have wanted to follow Church-teaching and receive 'ordinary' care.

Despite ignorant talk about 'brain failure' and insensitive talk about her being 'a vegetable' and 'as good as dead', her heart kept beating, her lungs kept breathing, all her bodily functions continued, all without 'life-support' or like assistance. With basic nursing care she might have lived for years. No-one seriously suggested that her assisted feeding was not working: the problem, from some people's point of view, was that it was working all too well! Nor was her assisted feeding a burden to her, physically or psychologically: she was unaware of it. Nor was it a great financial or logistical difficulty: her nursing care was covered by a trust fund, her feeding cost less than the average American spends on food and water, and it could easily have been administered at home by a family member.

The issue came to courts, legislatures and governments, and a media circus ensued. Ronald Cranford, a strong advocate of euthanasia and assisted suicide, described her case as 'the highlight of my career.' The courts directed that her feeding

Bishop Anthony Fisher, O.P.
John Paul II Institute for Marriage and Family, Melbourne;
University of Notre Dame, Sydney
Email: afisher@sydney.catholic.org.au

C. Tollefsen (ed.), *Artificial Nutrition and Hydration:
The New Catholic Debate*, 3–37. © Springer 2008

tube be removed and with it all nutrition and hydration; after some dispute about whether she could still swallow, the court also forbade spoon feeding and water by mouth. Her parents and priest were allowed only restricted access and were prevented from being with her at the end. On March 31 2005, just after nine in the morning, Theresa Marie Schindler Schiavo died, aged 41.

Terri Schiavo joined a string of prominent cases in the United States, Britain and Australia over whether to withdraw assisted feeding and hydration.[2] Because they demonstrate the deep philosophical divides in contemporary society—the so-called 'culture wars'—these cases were given a high profile. Post-modern cultures are intensely conflicted over issues of human nature, life and death, dignity and rights, relationships and social responsibilities. Even the most basic ideas of all— such as that there is a way things are; that things have a *nature*, essence or intrinsic way of being; that there is good and evil and that certain courses of action might be universally right or wrong or objectively so in a particular case—are hotly disputed. Fundamental differences which might once have been recognized as arguments in metaphysics, ethics and religion are now played out in hospitals, courts and the media, without yielding much insight. As a result underlying conflicts of values and beliefs are often left unidentified and people talk at cross-purposes, each assuming the other is homicidal or vitalist, authoritarian or uncompassionate.

Why the current enthusiasm of some health authorities and providers, public guardians and ethicists for withdrawing assisted nutrition and hydration from those at a very low ebb—ideally with the patient's prior consent through an advance directive, or at least with their family's consensus, but if needs be by force of law? Diverse motives converge here. One is euthanasist: by the time people need long-term assistance with nutrition and hydration they are presumed to be better off dead. There are also economic and logistical motives, as people conclude that such patients are not deserving of finite health resources and other energies. Also at play here is genuine concern for the freedom of patients (and their guardians) to say no to being over-treated now or in the future, and perhaps to being treated at all. All recognize the importance of being able to deliberate in healthcare on the basis of appropriate information and choose for oneself. All too often, however, this concern for freedom has become in today's world an *idolatry of the will*.

This idolatry is in fact one of the permanent possibilities in human culture and philosophy. From time to time in human history, the careful balance between practical rationality and strong will clarified by the great ancient and medieval thinkers is sacrificed, almost always in favour of will. Early twentieth century existentialism, 1930s and '40s fascism, mid-century consumerism, the sexual revolution of the 1960s and '70s, the 1980s and '90s triumph of democratic individualism over tyrannical communism—all these strands of twentieth century experience made the late twentieth century just such a time. In health and aged care it has meant a shift from professional paternalism to an absolutist notion of patient autonomy. Also in many other areas of life today individual will trumps reason and community. Where there is consensus that will rules, this is usually because there are strong wills behind it. It *suits* governments, health insurers, medibusiness, taxpayers and consumers to equate being human with having a will: for long-drawn-out and costly care of others is expensive for those providing them. As a result those least able to press their needs are most likely to have them dishonoured. If the test of a civilization is how it treats its weakest and most vulnerable members, cases like that of Terri Schiavo are emblematic for so much more than just the care appropriate to one particular person suffering a cognitive impairment.

## 1.1.2 Autonomy Talk

Bioethical talk of 'autonomy' is code for a multi-faceted concept carrying within it a mini-history of philosophical speculation on anthropology, social life and ethics. In classical philosophy the concept was essentially a political one: the rational control which by nature the soul has or shall have over human choice and welfare indicates the sort of control that legitimate authorities should assume in the state (Plato, *Republic* IV) and the sort of responsibility for understanding, accepting, making and improving the law which citizens must assume (Aristotle, *Politics* 3, George, 1993, Ch. 1). Christian thinkers (e.g. Augustine, Gregory of Nyssa and Aquinas) baptized this tradition with further reflection on the reflexive or self-creative effects of choice, on freedom as the liberty not merely to choose but to choose according to God's will and one's ultimate happiness, and on the complex relationship between divine providence and human (secondary) causality. For Kant autonomy was the state of will of one who recognizes that the demands of moral law are neither externally legislated nor subjectively constructed but are objective norms of reason clear to anyone with a well-functioning mind and a will to apply them.

In the classical tradition, at least up to Kant, *rational autonomy* might be said to be the capacity to make reasoned decisions for oneself—which implies not simply choosing and acting as we desire to, or even as we have reason to, but choosing and acting on reasons we endorse. But for most of the sons of the 'Enlightenment' (e.g. Hume and Mill) autonomy was not grasping the norms of reason but approving a course of action and experiencing the freedom to pursue it. Where autonomy was not doubted altogether (as in the determinist and social constructionist accounts of some Marxists, Darwinists and Freudians), many late 19th and early 20th century thinkers further disconnected freedom from nature and objective rationality, and tended to reduce autonomy to the raw pursuit of preference accompanied by 'authenticity' or 'conscience' in that pursuit (e.g. the Nietzscheans and the Existentialists). Remnants of some or all of these different concepts of autonomy—and in particular, *control, rationality*, and *preference*—are today intertwined in ever more elaborate and confused ways in popular conceptions of what it is to be human and to live 'a good life.'

While a student in Oxford in the early 1990s I attended a seminar given by Professor Ronald Dworkin. Week by week we went through the manuscript of a book Dworkin was writing, which was eventually to appear as *Life's Dominion* (Dworkin, 1993). The book was in large part an attempt to dissuade the US Supreme Court from pulling back on *Roe v Wade* in its then-forthcoming judgment in the *Casey* Case, as it was thought to have done in *Webster*.[3] In due course Dworkin hoped the court would use its legislative power to legalize euthanasia as well. The class was set up as a gentlemanly debate between Dworkin (with evolving text) and another Oxford philosopher, Bernard Williams, before an audience of adoring students. The class was swept along with Dworkin's high rhetoric about the dignity of the person and the sanctity of life and how respect for both required us regularly to kill un-born babies. There were hardly any expressions of dissent, with Dworkin's sparring partner really only helping with fine-tuning.

Finally Dworkin came to his draft chapter on euthanasia.[4] Till then autonomy was the trump card in all moral issues—the autonomy of adults anyway. 'But what about granny,' he asked, 'sitting around all day watching cartoons on TV and eating peanut-butter and jelly sandwiches?' The class laughed. 'She thinks she is happy,' he said. 'She wants her life to carry on. But we know better don't we? We know her life is now like the "white noise" on the TV after the station has stopped transmitting. We know she would be better off dead. And we know we should help that to happen.'

You could have heard a pin drop. A class full of autonomy-as-trumpets sud-denly saw where it might be leading. 'This is not really about autonomy at all, but about quality of life,' one till-now-wide-eyed student complained. 'No,' responded Dworkin, 'it *is* about respecting dignity and autonomy. When those are compro-mised by old age or disease we are better off dead. We know that for ourselves, so we should show simple charity to others . . .' For once the students were not happy with Dworkin. That gave me my opening, as a subversive amongst them, to raise a few questions in and out of class! When the book appeared in due course, that chapter had been considerably toned down. Nonetheless its *autonomy-personism* remains: the view that people who lack full autonomy are not fully persons. Sentimentality, nostalgia, speciesism or some other irrationality could distort our judgment here, of course, but on this account the description of Terri Schiavo by her guardian's lawyer as a 'houseplant'[5] was in some ways more accurate than her parents' reference to her as 'our daughter.'

Dworkin is one champion and Peter Singer another of bioethics descending from John Stuart Mill. Liberals advance at least two, not entirely compatible,[6] reasons for not feeding those who cannot exercise rational autonomy. The first is the view (sup-posedly from Mill's *On Liberty*) that what really matters in life is 'doing it my way': being able to pick and revise my own values and life-plans, make my own choices, satisfying my own preferences. The second view (after Mill's *Utilitarianism*) is that what matters in life is maximizing good sensations or fulfilled preferences and min-imizing bad or unfulfilled ones. Both views accommodate rationality to preference rather than *vice versa*. Both conclude that real respect for those who will never again be free choosers and for those around them who can still have sensations and preferences requires withdrawing any life-sustaining care and even, more actively, hurrying up their deaths. Both are powerful strands of contemporary culture. Their combined effect has been not only that 'autonomy trumps all' other moral concerns in rhetoric and practice, but that the kind of autonomy-that-trumps-all is no longer rational-critical endorsement of reasons (as in the classical tradition) and much more something like the freedom to act on immediate desires and sensations. Appeal to autonomy has slipped (without much opposition) from appeal to my reason to appeal to my wants.

In this paper I will adopt a broad definition of autonomy. I will ask why people who lack the exercise of rational autonomy in any or all of the senses I have indicated still matter, objectively speaking, and matter very much, and what the implications of this might be for how we treat them. I will take it, too, that rational autonomy, at least in the sense of being in control and acting for reasons we identify with and endorse, matters very much to us (subjectively). Most of us take the possession and frequent exercise of rational autonomy for granted; we abhor the prospect of its future reduction; and we miss it deeply when it is compromised in ourselves or someone we care about. Rational free beings naturally want to control their choices and destinies by critical reflection on their options and by taking decisions they themselves believe to be reasonable in the light of their circumstances. It is distressing to think of having to live, possibly for years on end, without being able to exercise rational autonomy. What is also perplexing, however, is the recent suggestion that we can or should move from the claim that 'rational autonomy matters very much' to the claim that 'people *only matter because* of their rational autonomy' and therefore to the conclusions that 'people who are unlikely ever again to exercise rational autonomy are *better off dead*' and that 'such people may be *killed* by action or neglect of basic care.'

## 1.1.3 Catholicism as a Sign of Contradiction

On these matters as on so many others Catholicism is a sign of contradiction. Why are some Catholics and others so concerned to protect the right of those who cannot exercise autonomy to be fed and watered? *Good* reasons here include a different conception from the liberal one of the human person and of respect for persons, of what is treatment, what is appropriate treatment, and what other kinds of care are appropriate. Amongst the multitude of philosophical anthropologies available today (person as functional system, as ghost, as rat, as computer, as sentient consciousness, as self-consciousness, as language user, as chooser . . .), Christians hold to an objectivist account of the person as a being that is material, living, animal, rational, free, social, emotional and immortal, and they offer metaphysical and biological arguments for this personhood from the first moment of the being's existence. This provides a clear and egalitarian ontological account of persons—including that of persons unable to respond appropriately to stimuli, think clearly or express themselves—and from this account certain norms of appropriate conduct towards persons may follow.

Entering this contemporary fray, Pope John Paul II in many of his speeches and writings, including *Veritatis Splendor*, argued that autonomy is *not* the source of human value or of human values; the value of freedom and conscience is in its pursuit and choice of objective truth; freedom is always freedom *in*, not *from*, the truth.[7] He re-emphasized the moral absolute against killing known to natural reason and confirmed by Christ and the Catholic tradition up to and since the Second Vatican Council (John Paul II, 1995, nos. 51 & 80). He repeated the long Christian 'call to an attentive love which protects and promotes the life of one's neighbour' (John Paul II, 1995, no. 15 *cf.* no. 13). John Paul II also expressed concern for the sorts of persons, professionals, institutions and communities we become when we fail to respect such precepts. In what some have labelled *Veritatis Splendor Part 2*, the encyclical *Evangelium Vitæ*, John Paul criticized the growing tendency 'to value life only to the extent that it brings pleasure and well-being', to view all suffering as 'an unbearable setback, something from which one must be freed at all costs' and to view 'the growing number of elderly and disabled people as intolerable and too burdensome.'[8] 'In this context,' the Pope noted, 'the temptation grows to take control of death and bring it about before its time, "gently" ending another's life.' He went on to argue that while this might seem logical and compassionate, it is in fact 'senseless and inhumane.' It is 'the height of arbitrariness and injustice' to take it upon ourselves to judge 'who ought to live and who ought to die' (John Paul II, 1995, no. 66). And it is an abuse of the notions of individual freedom and human rights to suggest that people have a right or a duty to die (John Paul II, 1995, no. 18).

John Paul's clarifications of the motives of euthanasia (whether selfish or compassionate) and the intention of euthanasia (to relieve suffering by deliberately hastening death) served to highlight the risks of a mentality which declares that some people have a 'virtually sub-human' quality of life, which regards their life as of 'no benefit,' and which accordingly judges their death as 'no loss.' It also served to highlight that euthanasia can be committed by omission of due care (so-called 'passive' euthanasia), as readily as by action such as poisoning. Thus he repeated Christ's injunction to feed those in need:

> In our service of charity, we must be inspired and distinguished by a specific attitude: we must care for the other as a person for whom God has made us responsible. As disciples of Jesus, we are called to become neighbours to everyone (cf. Lk 10:29–37), and to show special favour to those who are poorest, most alone and most in need. In helping the hungry, the thirsty, the foreigner, the naked, the sick, the imprisoned—as well as the child in the womb and the old person who is suffering or near death—we have the opportunity to serve Jesus. He himself said: "As you did it to one of the least of these my brethren, you did it to me" (Mt 25:40) . . . Where life is involved, the service of charity must be profoundly consistent. It cannot tolerate bias and discrimination, for human life is sacred and inviolable at every stage and in every situation; it is an indivisible good. We need then to show care for all life and for the life of everyone. Indeed, at an even deeper level, we need to go to the very roots of life and love (John Paul II, 1995, no. 87).[9]

In his subsequent allocution *On Feeding Those in a 'Persistent Vegetative State'* (2004), John Paul II made it clear that a refusal to feed unconscious people who are in need of food is an example of this kind of 'intolerable discrimination' and can be effective euthanasia.

I will return to *Evangelium Vitæ* and the *Allocution* later in this paper. Suffice it here to note, however, that the central concerns of the two documents are precisely what are at issue in contemporary debates over the withdrawal of feeding. This positions the Catholic *magisterium* very much at odds with some contemporary philosophers, theologians, lawmakers and health professionals. The views that patients suffering 'PVS' or like diminishments are dead, as good as dead or better off dead and/or that others can attribute to such patients the desire to be dead and/or

a picture of *human* welfare and so, without knowing it, they have already imported some objectivist notion of human *nature* even as they deny the egalitarianism of the human dignity idea. One question one might put to them is: if we could introduce genes or drugs or tissues to restore brain function (and so the exercise of rational autonomy) to a patient living in the 'limbo' of unconsciousness, would we regard this as correcting a handicap or creating a new being; as therapy or conception? Would we use the same name for the patient before and after the procedure? Clearly, we would be *curing* here, and there would be a someone we cured: we would not be creating or transubstantiating. We would be responding to the needs of a damaged person, restoring a function appropriate to a person's nature.

Alasdair MacIntyre argues against the élitist account of human personhood by arguing very persuasively that we are fundamentally *dependent* beings and that this dependence is not just chronologically but logically and ontologically prior to our independence or autonomy (MacIntyre, 1999). Reflecting upon our nature as physical, biological beings, MacIntyre also argues that our ethics must be grounded in our bodily, animal nature. Here he joins Aristotle, Aquinas,[12] John Paul II,[13] and many contemporary authors[14] who have argued that the body and bodily life are *not* merely instruments somehow distinct from and serving 'the real me', 'the self' or, as Richard Rorty calls it, my 'mind-stuff' (Rorty, 1991). All these writers demonstrate that mind-stuff approaches not only necessarily presuppose an indefensibly dualist conception of the human person but also adopt a radically denatured anthropology. Human beings are rational *animals*, living organisms, not angels or spirits connected or disconnected in some way to an animal body. They are their bodies. Their life *is* bodily life. Deliberately to end the bodily life of a human being is to kill that person.

## 1.2.2 Inviolability

Belief in and arguments for the equal inviolability of human persons often come paired with belief in and arguments for the equal dignity of every human being. The denial of one often comes with a denial of the other, but not always. Some concede that those lacking rational autonomy are persons with human dignity (whatever that might mean) but assert they are not inviolable. Humanity is divided, then, not into 'persons' and 'non-persons' but into 'protected persons' and 'unprotected persons', or into 'persons who should live' and 'persons who should not.'[15]

Historically, this move has been made by many of the same thinkers and legislators who made the élitist distinction between human and subhuman (examined in the previous section) or by others in the same communities. The results have included slavery, totalitarianism, apartheid, segregation, genocide... Any suggestion that modern societies like ours are of this sort would be met with howls of protest which, to some extent, would be justified. But before we get too smug, we should recall that no society before ours has dreamed of killing a quarter of its new members (before birth) or doing such killing so unashamedly. Asylum seekers and suspected terrorists are examples of other groups increasingly denied full human rights. Older people are progressively being pressed to sign 'living wills' that direct that they not be treated if they become incompetent (Tonti-Filippini, Fleming, & Walsh, 2004). Now it is proposed that a new class of human beings will be denied food and water, allowed to die of malnutrition or dehydration, and the bystanders will be told by law and medical ethics that that's alright. We may not be as morally superior to other élitist cultures as we imagine.

For the healthy to violate or neglect so openly those who suffer (whether against their wills, without their wills, or by forming their wills so that they themselves condone the violation) is a new form of disrespect for humanity. We are generally more careful today not to use the language of 'moron' and *'untermensch'*, 'subhuman' or 'inhuman', as some societies have done in the past, though cases such as Schiavo's often show our élitist petticoats. What we share with those societies we so readily deplore is the reversal of the proper protectiveness of the strong for the weak.[16] The Schiavo case demonstrates that in contemporary Western 'civilization' sickness, which always meant vulnerability, now means an even greater vulnerability: susceptibility not just to infection and injury but also to action and neglect aimed at worsening the condition, even to the point of destroying the sick person, *precisely because of sickness.* We literally add insult to injury (*Job*, Ch. 2),[17] as sickness launches the patient onto the slope of diminished 'personhood' and therefore diminished entitlements.

Those who argue for personhood but against equal protection and care for all persons (including inviolability) often adopt various kinds of camouflage rhetoric, such as that non-treatment is 'in the patient's own best interests', 'avoiding unnecessary suffering', 'easing along the inevitable' or 'letting nature take its course.'[18] But one suspects that what is often at issue it is not so much respect for the cognitively impaired person, but for the more autonomous, more powerful and more vocal bystanders. Sometimes it is asserted of the incompetent patient that 'she would not have wanted to go on in this way' and so 'she would not have wanted to be fed artificially.' Hypotheticals about what someone 'would have wanted' are dubious at the best of times, let alone when those constructing them are exhausted by the person's ongoing care. Even had a person made their preferences known in advance—which was *not* the case in the Schiavo, Bland or BWV cases—we might question the validity of such 'directives' made when a person does not yet suffer from the condition later being addressed, does not really know all the options in the situation which later arises, and so could not really make an informed decision.

Moreover, encouraging people to consent in advance to being starved should they become incompetent will not simplify the situation, ethically speaking. People, especially weak people, who respond to the invitation of authoritative persons to renounce their inviolability, are not thereby exercising true autonomy. Homicidal acts or omissions do not become right simply by becoming policies or by getting the victims to sign their own death warrants. In many ways that only aggravates the evil being perpetrated.[19] Protecting the norm of the inviolability of every human being, like protecting the egalitarian principle of universal human personhood and dignity, is as essential in these 'enlightened' times as in any dim dark age at which we wag our modern fingers.

But is the life of every human being *really* to be regarded as inviolable? Various cases have been made for this moral claim. One begins with the idea that *life* is a good basic to human choice and flourishing (John Paul II, 1995, no. 68). Life here signifies *organic* or bodily existence, its preservation, prolongation and transmission. Terms such as 'liveliness', 'vivacity', 'business life', 'social life' and 'the good life' depend for their sense upon the prior organic understanding of life. Some deny that organic life, as such, is a dimension of human flourishing or a good rationally pursued in human choices: life, they say, is only 'worth living' because of the things it enables. People only want to 'go on living' so as to be able to do their work, care for their children, enjoy playing tennis, and so on. On this view it is *biography* not life per se that matters: what we consciously experience and chose, what we write with our lives. On this account organic life a very important, *instrumental* good; but when it no longer serves other goods through conscious experience, choices and actions, it is no longer valuable.

To which one might respond: the instrumental uses of organic life do not exhaustively explain the value of life. Life is also enjoyed 'for its own sake.' That is why no-one expects us to give reasons for promoting life, avoiding death, and so on: we regard these objects as sufficient reasons *in themselves;* we value human life *per se;* someone can say meaningfully 'it's good to be alive' without having to explain what they are is doing with that life. It is this sense that life is *intrinsically* valuable which is behind our talk of human life as inviolable, of 'the right to life', and so forth. We express this same insight in many of our actions and institutions. We celebrate births and birthdays, and grieve over deaths and anniversaries of deaths. We delight in good health and recovery of health, and lament sickness, disability and pain. We send a congratulations card at a birth, a get well card at sickness, a sympathy card at a death. We bring children into the world and nurture them. We protect life through life-savers, sea-rescue operations, road safety laws, anti-smoking campaigns. We punish attacks on life through our legal systems. We promote life through our hunting and gathering, fishing and farming; through markets, food shops and pharmacies; through our health, welfare and education systems. In explaining what we are up to in this broad range of activities, communications, projects and commitments *life* is sufficient explanation. Bodily life is manifestly a good thing to have *in itself*, and not merely for the uses to which it might be put. Directly to deprive a person of life, therefore, is *prima facie* to deny him a good that is properly his and always to do him a harm.

A few qualifications are in order here. To say that life is a basic human good is not to say that the prolongation (and transmission) of life is reasonably to be pursued by all people, at all times and in all circumstances, at whatever cost to themselves and others, and by whatever means. Neither does it even mean that life is the only value or the most important value or an absolute value. Rather it says that the pursuit of life makes some human choices, activities, claims and commitments intelligible, whether or not they are reasonable in the circumstances 'all things considered.' Grisez and Finnis have argued convincingly that no one basic good properly overrides all others. There will often be good reasons to do things which protect or prolong life: the good of life itself, the other good things life enables, responsibilities to others, especially dependents, and so on. But there may well be concurrent good reasons *not* to do so: great burdens of various kinds for the person whose life would be prolonged or for those who would be engaged in their care, the risks involved and opportunity costs, etc.

To say that life is a basic good does not of course mean that life will be equally valued by each person or equally emphasized in their biography. Nor does it mean that every person will appreciate the value of life, i.e. acknowledge it steadily and explicitly as a reason for action. Some people obviously do not, or at least behave as if they do not. Many people report only really valuing their life after there has been some threat to or diminishment of it. Both rational clarity and feelings about matters of logical, empirical or practical facts may vary without impinging upon the truth of those matters. If this were not so, we could never judge whether our thoughts are true and our feelings appropriate to the circumstances.

Furthermore, to say that life is a basic good is *not* to say that we tend to desire it by itself. People don't just want to live but to live well, exercising amongst other things their rational autonomy in choices that contribute to their happiness and that of others. On the face of it this might seem to suggest once again that life is merely an instrumental good; however, one would not expect people to want to participate in any single basic good, abstracted from the range of ends which together constitute true flourishing. This is so precisely because each good is only one among the complex of basic goods which constitutes flourishing: truth, beauty, work, play—none of these goods would be appealing by itself in an existence (were that possible) deprived of all other goods. No one wants to suffer 'PVS' or dementia and go on living for long periods the subject only of the goods of (continuing) life, (diminished) health and (received) love, none experienced consciously. No one wants to see anyone else living like that. Nonetheless, even the severely cognitively impaired are living human beings: their life is their very reality as persons and as such remains a good, even if it is not consciously enjoyed by them and however little it appeals to us. Their death is a harm which diminishes them and the human community, even if it is also welcomed, at least by some. This explains why we still care for such people: such care ensures their continued participation in whatever goods of which they can still be subjects, maintains our bonds of interpersonal communion or solidarity with them, and expresses our benevolence and respect for them as living, if profoundly handicapped, members of the human family. It also explains why we do not harm such people, why we do not kill them or bury them alive, why we do not exploit them by live organ-harvesting or experimentation, why we do not sexually abuse them, throw them on the garbage heap, or otherwise subject them to indignity.

The 'inviolability of human life' doctrine follows from understanding that life is a basic human good, that good is always to be pursued and evil avoided, and that evil may not be done even to achieve some other good (such as 'merciful release'). The 'natural law' precepts to *preserve life* and *not to take human life* are thus determinations of the basic good of life and, like the good itself, are both underived and self-evident. The negative norm is held by the Catholic Church (amongst others) to be a moral absolute.[20] Of course, certain further principles, virtues and life-plans will be required if a basic good such as life is to be pursued reasonably and basic principles such as 'preserve life' and 'do not kill' are to be understood and applied appropriately to specific situations.

It is one thing to say that there is a positive norm requiring us to feed (ourselves, those in our care and those in our reach): it is another to assert that a failure to feed is always wrong. No-one can ever exhaust the demands of such a positive norm and we are not morally responsible for the deaths of every person we might conceivably have helped (e.g. by sending all our money to feed the starving in the Third World), as long as we are devoting our time and energies to other morally reasonable purposes, fulfilling our responsibilities. Even in particular cases of persons in need of food who are within our reach, there may be good reasons for not satisfying the positive norm, e.g. because to feed is impossible, ineffective, overly-burdensome to the patient or to others or because to feed conflicts with another (equally serious) duty to that person or to others. None of these failures to feed is homicidal. On the other hand, there are many situations in healthcare (as elsewhere in life) where one may choose to withhold or withdraw some kind of necessary care precisely so as to hasten death. The commonplace practice of denying handicapped infants even fairly simple surgical interventions to correct blockages (tagged euphemistically 'benign neglect by physician') is often, morally speaking, equivalent to active and avowed euthanasia. Though the agent may plead 'I didn't *do* anything', that is precisely the problem: he/she could have and should have done something but failed to do so because the agent thought the patient 'better off dead.' Whatever the legal situation, from the moral point of view it makes no difference whether one kills by action or omission of reasonable care, if killing is the goal. In such cases, the failure to fulfill the positive norm *to feed* is also a failure to observe the negative norm *against killing*.[21]

In really hard cases, such as those of people living in 'PVS', sympathy can tempt us to compromise such moral norms or to seek to make an 'exception', while we tell ourselves we can still hold the line 'as a general rule.' Rational reflection, however, as well as human experience suggest that the implications of such exceptions go far wider than the relief of particular hard cases. Here we bump up against the ultimate question for end-of-life ethics and indeed for all ethics: the mystery of evil. How are we to face ineradicable suffering, when we have tried all we reasonably can to combat pain, disease and dying? The pervasive temptation for modern man is to demand an immediate technological, consumer or government 'fix' and rail like a petulant child until that happens; and when a fix is impossible, this same consumer culture stands in gaping incomprehension, it goes into denial, withdraws its support, and/or marginalizes those who cannot be fixed so that the rest can carry on undisturbed. The fact is that there are evils we cannot 'solve' in any simple, morally acceptable way. Such evils can, however, call forth much that is most noble in the human spirit: patient endurance, perseverance, fortitude, even heroism on the part of patients, doctors, families and communities. Sometimes more *patience* will be asked of the bystanders than the *patients* themselves—and impatience will often be at the heart of the decision to stop feeding. As Benedict XVI put it in his homily at the Funeral Mass of Pope John Paul II: 'The world is redeemed by the patience of God. It is destroyed by the impatience of man.'

## 1.2.3 Feeding

If we do accept that those who lack rational autonomy are persons and should not be violated, is it clear that they should be fed and, if so, how and by whom? Gratian, quoting the Fathers, and the Second Vatican Council quoting Gratian, said 'Feed the man who is starving: for if you do not feed him you are killing him' (Vatican Council II, 1965, no. 69).[22] Starving or dehydrating someone to death by obstructing them from obtaining food and water or by failing to provide it when they depend on others to receive it, has always been considered not only killing but a particularly egregious form of killing. Why is that? I think it is because food and water are not only sources but symbols of life and community. To deny things which are so basic to someone is not only to deny them a need but necessarily undermines and ultimately denies all solidarity with them. When tyrants have starved people such as St Maximillian Kolbe to death, they have demonstrated contempt not only for the inviolability of life but also for the humanity and dignity of the person so painfully and degradingly killed.[23] Just as some actions are considered 'treasons' in addition to whatever wrong there is in the action itself, because they offend against the crown, the nation, or the community, so starving someone to death might be considered a treason against life.

Refusing food or water to someone is a powerful symbol of exclusion from the circle of the community, of humanity: but to do this when someone is weak or dying is especially revealing. Our normal response to frailty and dying is to shield those in this situation from the eyes of strangers, from lethal neglect of basic care and from all attacks—whether by TV cameras, greedy relatives, hospital number-crunchers, organ harvesters or others. With the frail and dying, protecting what is left of their lives and ensuring they have rudimentary care such as nutrition, hydration, warmth, prayer and company may be about all we *can* do for them. This shielding care maintains our solidarity with them while we still can and it declares to them, to ourselves and to society that these people still matter and matter very much. On traditional understandings, to ask 'should we care for the frail and dying?' is only a question for the callous, the invincibly selfish or those with no moral sense at all. This is because we apprehend that it is the same as asking: 'should we care for those who most need our care?' Caring for the frail and the dying is *criterial for caring*: it is part of how we understand, and how we show that we understand, what care *is*.

## 1.2.4 Assisted Feeding and Hydrating

Traditionally in healthcare feeding and hydrating have been regarded as nursing care rather than medical treatment. Philosophers since Aristotle and Hippocrates have noted that while certain foods and drinks, taken in appropriate circumstances, have medicinal properties, the primary purpose of food and liquid is not addressing any sickness but simply nourishing and hydrating the body so that the most fundamental processes underlying physical and mental health can occur. Thus *doctors* do not usually feed patients: nurses, family or the patients themselves do, and they do so whether they are sick or well. Not only are there different 'workforces' involved but different goals: while doctors focus on 'treatments' which seek to prolong life, cure disease, heal damage or halt degeneration, nurses (and others who do nursing) provide 'care' which sustains life in the meantime. Such 'nursing' care continues to be appropriate even when medicine has achieved a cure or when there is no more that medicine can do. Of course there may be some overlap and some of the same principles apply to deciding what should be provided to whom, but there may also be some differences of ethos and ethic between 'treatment' and 'care.'

There is, however, considerable dispute about how assisted feeding should be regarded. While inserting the tube, monitoring it and prescribing dietary supplements are usually performed by physicians, the actual feeding through the tube is usually performed by nurses, family members or, where conscious by the patient. The goal of the activity of the doctors, nurses, patients and family is a non-medical one, that of nutrition and hydration. Thus Pope John Paul II joined others in resisting the designation of assisted feeding as 'medical treatment.'[24] To recategorize food and water as treatments because doctors have given some assistance reduces medicine to nominalism and medical ethics to sophistry. Such relabelling may well function at an emotive rather than factual or ethical level as a pretext for the lethal withdrawal of food and hydration. These word games operate on the emotive rather than the objective factual level.

If there is a duty to feed (and hydrate), then there is a duty to feed a person what is appropriate to their needs and by means that are effective. Just as the positive duty to feed has its limits, so will the duty to assist feeding with various techniques and devices. The harder it becomes to achieve effective nutrition and hydration, the less comfort such provision also offers, the more burdensome the mode of delivery is to the recipient or to others, the closer the recipient is to death (and therefore to not needing nutrition), the less reason there will be to use such means. While such feeding would in principle be continued by spoon it should, if necessary, be performed by more complex means (such as a straw or feeding tube) if necessary. My thought, then, is that the case for feeding the frail, handicapped and dying, if needs be by tube, is based on two lines of argument: one, which focuses on the inviolability of life and the unethical euthanasist intentions of many denials of assisted nutrition and hydration; the other, which focuses upon the symbolic and social import of feeding and hydrating.

## 1.3 A Theological Account of Why Those Who Cannot Exercise Rational Autonomy Still Matter

### 1.3.1 From the Imago Dei to the Duty to Feed and Vice Versa

In my previous, philosophical, section I noted the common progress in 'pro-life' argument from the proposition that 'all human beings have human dignity' to 'the lives of those with human dignity are inviolable' to the conclusion 'there is a duty to feed (even artificially) all human beings.' There is a theological parallel which runs from the proposition that 'human beings are made in the image of God' to 'the lives of all human beings are sacred' ('the sanctity of life') to the conclusions that 'there is a duty to feed all human beings, if needs be artificially.' Once again these propositions are conclusions and shorthand expressions for some quite complex arguments, profound concepts built out of other yet more basic ones. To describe people as 'the image of God', for instance, is to quote an ancient and somewhat opaque scriptural text and to express a stance towards the questions of the divine-human relationship, human self-understanding, the limits and scope of human freedom, rationality and stewardship. To talk of 'the sanctity of life', at least in a theological context, is to identify human beings as created, redeemed and destined for greatness in this life and the next, to identify God as the Author and Redeemer of those lives and to assert that only He may give or take such life.[25]

I suspect that in theology (as indeed in philosophy) argument often works in a reverse direction to what first appears or that premises and conclusions qualify each other in complex ways. I suspect, for instance, that we sometimes reason, whether consciously or not, from the 'conclusion' (perhaps given by revelation or the magisterium or common sense or intuition) that we should feed those in need, if necessary with some artifice, to the 'premise' that all human life is sacred, not just the lives of the fit and active. Having grasped this, we elaborate a theology of man as 'the image of God', of life as being so precious as to be worthy of the tag 'sacred', of he norm of 'inviolability' and of the duty 'to feed.' I will begin this theological exploration therefore, with the precept to feed, even artificially.

### 1.3.2 "All I Ask as I am Dying is This: Honesty, Comfort and the Food I Need" (Prov 30:7–8): Church Documents

Many individual bishops and bishops' conferences around the world have addressed the issue of assisted feeding and hydration for people suffering from 'PVS' and like conditions over the past decade or so. In 1992 the US Bishops' Committee for Pro-Life Activities issued a pastoral statement *On Nutrition and Hydration,* concluding that there must be a presumption in favour of tube-feeding for persistently unconscious patients and that such measures should not be withdrawn unless they offer no reasonable hope of sustaining life or pose excessive risks or burdens. A similar position was taken by several state bishops' conferences (Florida, New Jersey, Washington-Oregon, and Pennsylvania—but not Texas). In 1994 the US Bishops approved a new set of *Ethical Directives for Catholic Health Care.* Here they repudiated two extremes: 'on the one hand, an insistence on useless or burdensome technology even when a patient may legitimately wish to forgo it and, on the other hand, the withdrawal of technology with the intention of causing death.' They agreed with previous statements that tube-feeding is not morally obligatory when it brings no comfort to a person who is imminently dying or when it cannot be assimilated by a person's body; however, they stated 'there should be a presumption in favour of providing nutrition and hydration to all patients, including patients who require medically assisted nutrition and hydration, as long as this is of sufficient benefit to outweigh the burdens involved to the patient' (National Conference of Catholic Bishops, 1994, nos. 56–58).

This position was confirmed by Pope John Paul II in his 1998 *Ad limina address to the bishops of California, Nevada and Hawaii*. He reminded the bishops that

> a great teaching effort is needed to clarify the substantive moral difference between discontinuing medical procedures that may be burdensome, dangerous or disproportionate to the expected outcome—what the *Catechism of the Catholic Church* calls "the refusal of 'over-zealous' treatment"—and taking away the ordinary means of preserving life, such as feeding, hydration and normal medical care. The statement of the United States Bishops' Pro-Life Committee, *Nutrition and Hydration: Moral and Pastoral Considerations*, rightly emphasizes that the omission of nutrition and hydration intended to cause a patient's death must be rejected and that, while giving careful consideration to all the factors involved, the presumption should be in favour of providing medically assisted nutrition and hydration to all patients who need them. To blur this distinction is to introduce a source of countless injustices and much additional anguish, affecting both those already suffering from ill health or the deterioration which comes with age, and their loved ones (no. 4).

In 2001 the Australian Bishops unanimously approved Catholic Health Australia's *Code of Ethical Standards*. Like the American code it proposed a strong presumption in favour of feeding, even artificially:

> Continuing to care for a patient is a fundamental way of respecting and remaining in solidarity with that person. When treatments are withheld or withdrawn because they are therapeutically futile or overly-burdensome, other forms of care such as appropriate feeding, hydration and treatment of infection, comfort care and hygiene should be continued. Nutrition and hydration should always be provided to patients unless they cannot be assimilated by a person's body, they do not sustain life, or their only mode of delivery imposes grave burdens on the patient or others. Such burdens to others do not normally arise in developed countries such as Australia (no. 5.12).

The Australian Code rehearsed the tradition that therapeutically futile and overly-burdensome treatments may legitimately be foregone, noting that:

> The benefits of treatment include preservation of life, maintenance or improvement of health, and relief of discomfort. They do not include deliberately shortening the life of a person who is sometimes wrongly described as "better off dead" ... The burdens of treatment to be properly taken into account may include pain, discomfort, loss of lucidity, breathlessness, extreme agitation, alienation, repugnance and cost to the patient. In some cases, the burdens of treatment may also include excessive demands on family, carers or healthcare resources. Judgments about the futility of a treatment outcome must be distinguished from judgments about the "futility of a person's life": the former are legitimate, the latter are not (Catholic Health Australia, 2001, nos. 1.13 & 1.14).

The first Vatican document specifically to address the matter of tube-feeding was the *Charter for Health Care Workers* published by the Pontifical Council for Health Care Workers in 1994. It restated the usual distinctions between homicidal withdrawals of care and those justified by imminence of death or burdensomeness. It then added a telling rider: 'The administration of food and liquids, even artificially, is part of the normal treatment always due patients when this is not burdensome for them: their undue suspension could amount to euthanasia in a proper sense' (no. 120).

The following year saw the publication of Pope John Paul II's great encyclical on bioethics, *Evangelium Vitæ*. There he condemned euthanasia as 'a grave violation of the law of God', symptomatic of the 'culture of death' in many Western countries and contrary both to natural law and revelation as mediated by the magisterium (John Paul II, 1995, no. 65). In elaborating his rich theological argument for this, the Pope drew on a long line of sources from the Old and New Testaments, through the Fathers and Scholastics, to the writings of his predecessors (especially Pius XII), the Second Vatican Council and the curia. He was careful to make the necessary distinctions between euthanasia ('an action or omission which of itself and by intention causes death, with the purpose of eliminating all suffering') and appropriate pain relief or non-treatment. Catholic teaching, he recognized, has never required the prolongation of life at all costs; 'heroic', 'extraordinary' or very burdensome treatments may properly be foregone, especially when death is clearly imminent and inevitable, 'so long as the normal care due to the sick person in similar cases is not interrupted.' Though what might be included amongst 'normal care' was not spelt out in *Evangelium Vitæ*, it had already been clarified a few months before as including 'feeding and hydration, if needs be artificially assisted' in the *Charter for Health Care Workers*.

In 1998, as we have seen, John Paul again insisted 'the presumption should be in favour of providing medically assisted nutrition and hydration to all patients who need them'. In his 2004 allocution *On the Care of those in a 'Vegetative State'* he clarified the matter even further. He insisted that people who are unresponsive to stimuli, demonstrate no awareness of self or environment and seem unable to interact with others, should not be demeaned by tags or behaviour that imply they are less than human. In the face of trends of thought demeaning the dignity of the person suffering 'PVS' he reaffirmed strongly:

> that the intrinsic value and personal dignity of every human being do not change, no matter what the concrete circumstances of his or her life. A man, even if seriously ill or disabled in the exercise of his highest functions, is and always will be a man, and he will never become a 'vegetable' or an 'animal.' Even our brothers and sisters who find themselves in the clinical condition of a "vegetative state" retain their human dignity in all its fullness. The loving gaze of God the Father continues to fall upon them, acknowledging them as his sons and daughters, especially in need of help (John Paul II, 2004, no. 3).

The 'fundamental good' of life, the Pope reminded us, is not outweighed by quality of life or cost considerations and positive measures must be taken to support such people and their loved ones.

> "Quality of life" considerations, often actually dictated by psychological, social and economic pressures, cannot take precedence over general principles. No evaluation of costs can outweigh the value of the fundamental good of human life. Moreover, to make decisions regarding a person's life on the basis of someone's external evaluation of its quality

amounts to attributing more or less dignity to that particular person, thus introducing into social relations a discriminatory and eugenic principle (John Paul II, 2004, no. 5).

In *Evangelium Vitæ* John Paul II had already written of the 'intolerable' neglect that some of the elderly, handicapped and dying experience even in affluent nations. He exhorted us "to preserve, or to re-establish where it has been lost, a sort of covenant between the generations", a relationship of acceptance and solidarity, closeness and service. His 2004 allocution might therefore be read as unpacking something of the implications of this covenant relationship by considering how we should regard and care for the persistently unconscious.[26]

A last series of statements which might be considered are those of several bishops and of the President of the Pontifical Academy for Life, Bishop Elio Sgreccia, against the court-ordered removal of Terri Schiavo's feeding.[27]

Thus instead of the 'Never Feed', 'Always Feed' and 'Seldom Feed' views proposed by some with respect to assisted feeding of people who are at a very low ebb, the *magisterium* has consistently proposed what I call the 'Usually Feed View' (Fisher, 2005) and repudiated both a 'vitalism' that would tube-feed even when this no longer works or works only at a grave burden to the patient or others, and a euthanasist approach would deny food when the patient is judged better off dead. Everyone is entitled at least to food and water, clothing, shelter, sanitation, company and prayer. So if they need help, even if it includes some artifice, with achieving nutrition or hydration or clothing or sanitation, and it can easily be given them, then it should normally be given. Even persons suffering 'PVS' or like conditions have the right to such 'basic', 'natural', 'normal' or 'minimal' care as John Paul II called assisted nutrition and hydration in these cases. As the Pope pointed out, this kind of care is (only) 'in principle' obligatory as long as it achieves its 'proper goal' of nourishing or comforting (John Paul II, 2004, no. 4). Thus the Catholic tradition, like the Hippocratic one, has long held such interventions inappropriate:

- where the patient has died;
- where the patient is imminently dying;
- where the delivery of such nutrition and hydration is ineffective in feeding, hydrating or comforting the patient ('futile');
- where the mode of delivery is too burdensome for the patient; or
- where the mode of delivery places an unreasonable burden upon others.

This means that, according to the Catholic magisterial tradition—and the natural law philosophy and theological sources upon which it is based—it will sometimes be appropriate to withhold, reduce or withdraw assisted nutrition and hydration. It also means that *prima facie* assisted nutrition and hydration should be given to those suffering 'PVS' or like conditions. Despite some ill-considered talk, they are not dead, not dying, not burdened by assisted feeding. Tube-feeding does work for them in the same way that it works for anyone else, sustaining their bodily life and thus their person and it is usually relatively easily and inexpensively provided, at least in the First World. The presumption in favour of tube-feeding, when this is necessary to sustain life, is thus well supported in the recent magisterial tradition.

Let us now examine whether there is a basis of the 'prior' claim that there is a duty to provide feeding and hydration.

### 1.3.3 "When I was Hungry Did You Feed Me?" (Mt 25:31ff)

What then are the basis, limits and scope of the duty to feed in the Catholic tradition? All religion is bound up in some way with food and feeding, feasting and fasting. Religion teaches the mystical, communitarian and symbolic dimensions of offering food, of eating, of gluttony and of moderation. There may be food taboos, fasting and abstinence on occasions, but feeding those in need of food is required in every serious religion and appears almost antiphonally as a charge in Judaism and Christianity.[28] Food and drink are God's good gifts to be shared in turn with others.[29] According to the prophets, only the knave and the fool 'lie about God, fail to feed the hungry or deprive the thirsty of drink' (Isa 32:6f.). No genuinely religious man can stand by and do nothing while another starves—in fact few irreligious people can comfortably do so. It is the very antithesis of religion that always feeds, succours and offers food sacrifices to its gods precisely so that the spiritual advantages of food are made more fully available to the hungry. Withholding food so as to kill contravenes religion, just as it does natural reason.

But what was and is different about the Christian religion here? The Scriptures single out an aspect of Jesus' ministry that marks his attitude to food (and so that of the Christian religion) as potentially different. In the Gospels we hear one of the more spiteful pieces of gossip about Jesus: he was nick-named 'glutton and a drunkard' (*Mt* 11:16ff; *Lk* 7:31ff; *cf. Mk* 2:18ff *par*). It was not just that Jesus hungered and thirsted like any human being and so ate and drank (*Mt* 4:2 *par*; *Mk* 2:23 *par*; 11:12; *Jn* 4:6f; 19:28). The complaint seems to have been that Jesus and the lads were too worldly by half; that genuinely religious people (like John the Baptist) would abstain, especially if the end-times were coming; but Jesus and the lads were 'party animals.' Elsewhere I have suggested that the complaint on the lips of the Scribes and Pharisees was even more sinister: it implied Jesus was a sluggard, a wastrel, useless eater, who would come to nothing and so deserved to be denied food, even killed (Fisher, 1992).[30] In the context of our present discussions, this is especially poignant.

Such a charge could not hope to stick unless Jesus was in fact an enthusiast for food and drink—and so it seems he was. The Gospels record many feeding stories involving Jesus: Jesus' relatives, especially his good Jewish mother, seek to ensure that his ministry does not get in the way of his having proper meals; Peter's mother-in-law, cured of fever, gets up immediately to serve supper; having raised Jarius' daughter, Jesus' first direction was to give her food; and his dear friends Martha and Mary squabble over serving the dinner (*Mt* 8:14f *par; Mk* 3:20f; 5:35ff; *Lk* 10:36ff. *cf. Jn* 4:31; 12:1ff). The Jesus of the Gospels was often at wedding feasts, Pharisees' dining tables, eating with tax collectors and sinners, 'at home' with his friends or out hosting picnics in the hills.

In fact the turning points of Jesus' life are always marked by eating and drinking. His first great sign is turning water into wine; his first preaching to a gentile began with a request for a drink and ended with a promise of endless living water; his most recorded miracle is the multiplication of loaves and fishes; his last wonder before his ascension is the huge haul of fish (*Mt* 14:13ff *par*; *Jn* 2:1ff; Chs. 4 & 21). All these miracles were of end-time proportions, divine in their extravagance, a foretaste of the longed-for messianic banquet (*Isa* 25:6–8; 1 *Sam* 2:5; *Lk* 1:53). As his ministry came to its climax, he took his closest friends aside for a last meal, investing the Passover Seder with new significance: his own Pasch memorialized and perpetuated in the Eucharist (*Mt* 26:20ff *par*; *cf. Jn* Chs. 6 and 13). Before returning to the Father he dined again with disoriented disciples in Emmaus, with confused apostles at the Sunday gathering and with his nearest and dearest at the lakeside breakfast in Galilee (*Lk* 24:13ff; *Mk* 16:14; *Jn* ch 21).

All this partying scandalized those who closed their ears and hearts to Jesus but excited those open to his teaching. So when he wanted to describe the kingdom of God or the afterlife or forgiveness or ministry or himself, time and again he chose images of food and drink, feasts and parties. He told parables about vineyards, grapes, wine and wineskins; about wheat, yeast and bread; about oil, mustard seeds, figs, mint, dill, cumin, eggs, fish and a fattened calf. He described prayer as asking Our Father for daily bread and forgiveness as a father holding a feast to celebrate his prodigal son's return (*Mt* 6:11 *par*; *Lk* 11:5ff; 15:11ff). Christian life is about bearing fruit and yielding a harvest (*Mt* 7:16 *par*; 12:33 *par*; 13:23 *par*; *Lk* 3:9; *Jn* 12:24; 15:5). Christian leaders should be wise stewards who feed their household at the proper time, shepherds who feed Christ's sheep, their preaching savoury like salt (*Mt* 5:13; 24:45ff; *Jn* ch 21; *cf. Lk* 16:1ff). The kingdom of God is like a wedding party, where Jesus' disciples eat and drink at his table (*Mt* 22:1ff; 25:1ff; *Lk* 14:15ff; 22:27–30). And how does Jesus describe himself and his mission? My food is to do the will of my Father (*Jn* 4:32ff). I am the bread of life (*Jn* ch 6). And how does he leave himself for us? As food: Jesus' body and blood, under the species of bread and wine, the staple foods of life. Jesus is remembered in the sacred meal, really present in the food and drink (*Mt* 26:26ff *par*; *Jn* 6:48ff; 1 *Cor* 11).

Elsewhere I have argued that attitudes to food and drink have implications that run deep for our theology of creation and eschatology, incarnation and redemption, sacramentality and spirituality, politics and ethics (Fisher, 1995). Here I want to suggest that our attitudes with respect to whom we feed and our 'practices' (in MacIntyre's [1984] sense) with respect to how we feed them say something powerful about both them and us. With respect to whom we feed, it tells something of who is the 'in-group' and who is on the outer described in 1.2.1 above. The theological parallel is *communion*. When we share food, especially the Eucharistic meal, we are 'in communion' and when we don't, it signifies some rupture. To refuse to feed or water someone, on this account, is not merely to fail in an ethical duty: it is to excommunicate them, to place them outside the pale of human friendship and deserving, to deny them the status of brother and sister, and to record some defect in fellowship. This is why Christ made whether we feed the starving a test of communion and ultimately of salvation: to refuse them food and drink is to refuse them fraternity and ultimately to refuse the God who made the needy and the 'little ones' our special responsibility (*Mt* 25:31–46).[31]

Jesus' test of salvation—did you feed the little ones?—also highlights an important aspect of the anthropology and sociology of eating: the ritualisation of power and service relationships. Where a meal is held and at whose initiative, who is invited or who excluded, how and when they arrive and depart, what they wear and bring, where they sit, what sort of food they are served and how, and the conversation: all these things signal not just the in-groups and out-groups but who is where in a particular social pile, the hierarchies and influences, the bonds of interdependence, approval, loyalty and nourishment in the broader sense. Such matters were obviously very much at issue in Jesus' eating and drinking: his wedding feast parables are full of concern about who is invited, who declines and who is refused admission or ejected after arrival, where people sit, what they wear, and so forth.[32] He praised the woman who washed his feet for dinner and criticized his host who had failed to do so; he criticized those who obsess about cleaning pots and hands before supper while leaving their hearts unprepared. His Last Supper is replete with such details: Jesus directs the place, preparation and course of the meal; he invites only his intimates; he washes their feet; he blesses, breaks and distributes the bread and blesses and passes the cup; he engages in highly significant conversation with them all but also with particular individuals. One lies against his breast, others near at hand or further away—even the traitor is offered food, before being dismissed.

Thus whom Jesus is recorded as having fed and with whom dined was highly significant. Jesus' critics complained not just that he took food and drink too seriously but also about the company he kept at table. He was, it seems, altogether too inclusive, bringing people into relationship with him rather than keeping them at a distance, regardless of whether they were ritually impure, morally dubious or socially outcast. Many of them were far from powerful, far from fully autonomous or beautiful or successful. Though none suffered from 'PVS', one at least was actually dead before he raised and fed her (*Mk* 5:35 ff)! Such inclusiveness was, of course, deliberate and it was subversive. So were the rôle reversals: as Mary predicted in her *Magnificat,* in God's kingdom the poor would be filled with good things while the rich would be 'sent empty away'; a prodigal son feasting while his law-abiding brother excommunicates himself; the high and mighty self-excluded from the wedding banquet while the tramps are dragged in from the highways and byways; the rich man in hell while the starving one goes to Abraham's bosom (*Lk* 1:53; 12:37; 14:15ff; 15:25ff; 16:19ff; 22:27). Jesus' feeding miracles undermined the system of public patronage and were quickly read as political. And finally, in the Eucharistic texts we find the most striking teaching of Christ about food or through food: here he *is* the food. A more comprehensive overturning of what we might call 'power eating' would be hard to imagine. Far from using the feast to exercise control, Jesus makes himself the waiter and the meal, washing bodies and feeding souls, emptying himself of all pretensions to power at the very moment when all authority is given him in heaven and on earth. His moment of glory would be precisely when his body would be broken and his blood spilt for the world—and at that moment he would once more join all those in need as he cries out from the Cross 'I thirst'(*Jn* 19:28 *par*; cf. Chs. 12 & 16).

It is hardly surprising that Jesus' food practices posed difficulties for the early Church: the question of table fellowship between Jewish and gentile Christians divided the churches in Galatia and of the Gospels; there were also divisions about who should be fed and how at the Corinthian Eucharist and in James' congregation, over idol meat in Corinth and Thyatira, and over hospitality for itinerant prophets in the Third Letter of John. Paul it was who first formulated the idea of unity-in-Christ in a way which prevented the power structures and ideologies of pagan antiquity from finding a foothold in the nascent churches; this unity-in-Christ was to be expressed in the sacred meal, so that the Eucharist would function as a social cement in the life of the community. However, Paul, James and others immediately recognised that there is something deeply artificial about Eucharistic egalitarianism while ever members of that community are in need of food or otherwise neglected. How, then, are we to respond?

"Now give them something to eat yourselves!" Jesus commanded (*Mt* 14:16). The Church must be the stomach with which Christ still feels *splangchnizomai*, that stomach-churning compassion he once felt for the hungry crowd which moved him to feed them (*Mt* 14:13 ff *par*). Here I think we come to the heart of what is described so clinically in our bioethics documents as 'the presumption in favour of feeding' and it is a 'presumption' at the heart of the mission of the Church. Christians are called to feel gut-wrenching pity for those starving physically or spiritually and to respond by feeding them what they need.[33] That was precisely what the early Christians did, taking up collections of food and money to distribute to the poor, even appointing specialists—the deacons—to ensure this happened (*Acts* 4:34f; *Acts* 6:1 ff; *Rom* 15:26f; 1 *Cor* 16:1 ff; 2 *Cor* 9:13). Social historian Rodney Stark has suggested that the catalyst for the spectacular take-off of Christianity in the Græco-Roman world was its special appeal to the poor and needy (Stark, 1996). With surprising speed Christianity overturned popular morality and social expectations all around the Mediterranean world, challenging and converting cultures as well as individuals.

"See how these Christians love each other!" people said in astonishment (Tertullian, *Apologia* 39).[34] The early Christians were notorious for their respect for every human person: they refused to engage in the commonplace practices of abortion, infanticide, suicide or euthanasia, even in hard cases; they looked after the poor, starving, widowed, crippled, sick, elderly and dying, even in difficult situations; they stayed around even when there was a plague. Their simple, egalitarian approach had a tremendous kerygmatic effect. No longer did racial, ethnic, cultural, citizenship, social or gender differences mark the boundaries of moral concern and obligation. Now people of every class and background were to be loved and protected for themselves. The distribution of food and other alms to the poor preached more powerfully than the words of Peter and Paul—and it preached the subversive inclusiveness of the nobodies in this new 'kingdom of God.'

In due course the Christian virtue of *hospitalitas* meant the erection of the first hospices, poor houses, soup kitchens and feeding stations, in local churches and monasteries and later the emergence of great social institutions such as Catholic hospitals for the poor, orphanages, chivalric orders of hospitallers, St Vincent de Paul conferences, Caritas and the rest. We now take all this for granted but without Matthew chapter 25 and a whole Bible full of charges to feed and give drink to those in need it would never have happened. In the process this ethical imperative changed not only how the starving were treated but also how they were viewed: even the lives of the least were sacred.[35]

The biblical significance of food and drink, eating and drinking, suggests that even where feeding has to be assisted it should be done in a way that is as close as possible to the experience of a communal meal. No one would pretend that PEG-feeding is as humanly satisfying as enjoying a several course meal with friends. It is a poor substitute for taste, texture and company. (So, of course, are many modern 'meals.') Though tube-feeding ensures some of the same values are achieved as ordinary feeding, it can also mark us off and separate us from others. There is a real challenge to humanise feeding especially in institutional environments, as indeed there is to humanize many other aspects of the care of the sick and dying, making it as familiar and close to the patient's 'normal' experience as possible.

### 1.3.4 The Sanctity of Life and the Imago Dei

Anthony Castle tells the story of a leper colony in the most heart-breaking sense of that term: where people lived with nothing to do except wait for death, a warehouse for lonely, abandoned, blank-faced people. Yet one man kept a gleam in his eye, could smile and express gratitude. The sister in charge was anxious to know the reason for this miracle: what kept him clinging to his life, his dignity, his humanity? She watched him for days and noticed that there used to appear above the high, forbidding wall, every day, a face. It was the tip of a woman's face, no bigger than a hand, but all smiles. The man would be there, waiting to receive the smile, this food of his strength. He would smile back, the head would disappear and then his long wait for the next day would begin afresh. The sister one day took them by surprise. He said simply: "She is my wife." After a pause he explained: "Before I came here, she hid me and looked after me with anything she could get from the native doctors. Every day she would smear me all over with their pastes—all except one tiny corner of my face—just enough to put her lips to. But it could not last. They picked me up and brought me here. So she followed me here and when she comes to see me every day, I know that it is because of her love that I can still go on living" (Castle, 1994).

Talk of *the sanctity of life* often functions as a kind of Christian version of the secular accolade 'the dignity of the person' and/or the philosophers' norm 'the inviolability of life.' Many of my comments in 1.2.1 and 1.2.2 might apply here. All too often 'sanctity of life' is contrasted with 'quality of life', as if Christians want longer, low-grade lives and seculars want better if shorter lives. Neither is true of course: most people, religious or not, want long, good-quality lives for themselves and those they care about. Where *quality-of-life* talk does mark a real difference is where some argue that those below a certain quality-of-life threshold do not command the same respect and care as those above it Here sanctity-of-life talk functions, much like dignity talk, to insist that all human lives—*being sacred*—equally deserve care and respect. In addition, much like inviolability talk, it also functions to insist that even low-quality lives should not be deliberately shortened. The problems with contemporary quality-of-life thinking are well known and I need not repeat the arguments. I would note, however, that this kind of thinking is far from the monopoly of the non-religious. Christian versions of the quality-of-life threshold regularly appear, e.g. in the better-dead-than-not-having-spiritual-experiences line which I will examine below.

One way of reading 'sacred' in 'the sanctity of life' is to say it means that Christians have a reverence for human beings in excess of that commanded by secular 'dignity' or that they take a more absolute line on the inviolability of human life than merely philosophical argument would warrant. I want to suggest that sanctity talk has most bite at the margins, i.e. when it is hardest to hold on to the principle of not killing the innocent. Take, for instance, the philosophers' chestnut of the man in the burning car who cannot be rescued and begs to be put out of his misery; or the (rather more common) example of a couple considering abortion because their child has anencephaly. Even those who hold to the infinite value and inviolability of the human person are tempted to make exceptions in such cases. The contrary tug in such people or in others, against killing even in such awful circumstances, could be plain superstition, stubbornness or insensitivity. But it might be due to something else and that something else is called 'sanctity of life.' If I am right, it is a quality in the valuer as much as in the person valued and I think it is told better in examples than in words: Frederick Ozanam, Mary Aikenhead, Catherine McAuley, Frances Xavier Cabrini, Damian of Molokai, Mother Teresa of Calcutta, Jean Vanier and Helen Prejean are all relatively recent examples. There are many more ancient examples of saints who saw Christ present in the most desperate of people and hung around to help.

An example from my own Dominican tradition is St Catherine of Siena. Her first biographer, Raymond of Capua, wrote of Catherine's 'street ministry' caring for the most hopeless cases, offering her gentle touch to lepers and worse, refusing to flee as others did in the face of bubonic plague. She was very aware of her impotence to do more than basic nursing, praying and loving. In 1375 a Sienese youth, Niccolò di Tuldo, was condemned to death for a political crime. Hearing of his bitterness and despair Catherine threw herself into accompanying him on death row. She built up his courage and persuaded him to receive the Last Rites. At his request she went all the way to the scaffold with him and ultimately caught his severed head in her hands. Like Our Lady at the foot of the Cross she knew there was no more she could do than stand by and pray. This standing-by is, however, precisely what I think is meant by reverence for 'the sanctity of life'—even when it is 'hopeless', even when there is 'nothing we can do', there *is* hope and there *is* something we can do: we can stand by, watching, praying, loving. These incidents in Catherine's life came to a head when, furious at the moral evil of the system in which young men died on the executioner's block and at the natural evil of a world in which young women and children died of plague, she went to remonstrate with Christ on the Cross. As was not uncommon with mediæval mystics, the corpus actually spoke back to her!

"Turn around," he said, "and see who it is I love enough to die for." And turning around she saw all the halt and the lame, as well as the privileged and the perfect, the victims and their persecutors, all of them. Her task, she knew, was to expand the range of those she loved and to persevere in her care for them, no matter how repugnant she found them at times (Raymond of Capua, 1960).

That ability *to reverence those we find repugnant* and *to care for those for whom we feel we can do nothing* is, I think, where sanctity of life talk really bites. It is here that the notion of the *imago Dei* also comes in. St Catherine, Mother Teresa and the others I have mentioned reported that they could (sometimes at least) see God in those they nursed. Catherine's talk about God being *pazzo d'amore,* drunk or insane with love for us and about our catching that wild love from him, cut in precisely as the condition of her lepers became hopeless, as their bodies and spirits disintegrated, as they faded out of consciousness or were dying, when all she could do was give them basic care and wait. It is in those very cases that I think the liberal account of why people matter and the Christian one most radically diverge. For Mill, Dworkin and Singer, human persons matter because of sentience and mobility, preferences, hopes and plans, reason and choices, language and social interaction—all the stuff of rational autonomy. For Christians, while *all those things matter*, what makes people *so valuable* is their creation in the very image of God, their creation as the kinds of beings who will ordinarily exercise not only rational autonomy but unitive love and other capacities. What matters is their restoration to God's likeness despite their brokenness by the redemptive sacrifice of Christ; their sharing human nature with a God who became man so that man might become god (*Catechism* no. 456 after St Athanasius); the graces they receive and enact in this life in good (not merely free) choices; and the destiny to which they are called in heaven. As MacIntyre and Hauerwas have both demonstrated, it is often the 'profoundly disabled' who best draw our attention to what we really value—or should really value—in the human person: namely, their intrinsic, metaphysical nobility rather than their presently apparent, contingent abilities (MacIntyre, 1999; Hauerwas, 1986).

What I am suggesting, then, is that the Christian notion of the sanctity of life is more than secular dignity and inviolability dressed up in religious poetry. With the eyes of faith one comes to see every human being, and especially those most desperate, repugnant or beyond help, as the image of God, as the suffering Christ, as a potential spiritual sibling and Temple of the Holy Spirit, worthy not just of 'respect and care' but of a 'mad' love akin in some ways to worship. Indeed the 'weaker', 'less respectable', 'low quality-of-life' people are precisely the parts of the Body of Christ for whom Christians are challenged to demonstrate particular sympathy and protectiveness.[36] On this account directly to kill or neglect-to-death an innocent human being involves more than the loss of that person and all he or she means to others, more than the harm it does the killer and the community—though it does indeed involve all those losses. Killing or abandoning someone to death is also, and ultimately, a kind of desecration of something/someone sacred, an attack upon the God of whom the victim is an icon. So says Paul, quoting the wisdom literature and recalling no doubt Jesus' command to love even your enemies: "if your enemy is hungry, feed him; if he is thirsty give him drink" (*Rom* 12:20; *Prov* 21:25).

That, I think, is why the early Church put murder with apostasy amongst the most horrendous crimes: not just because both are very bad (for there are many, very bad sins), but because to kill another human being was not merely to do an injustice but directly to attack the Author of Life, to usurp his rôle and so to sin against faith, hope and love. Talk of sanctity of life tries, however limpingly, to capture something of the reverence or awe that religion has before the mystery of life and death, and something of the shudder down the spine the Christian feels at the thought of killing or neglecting-to-death another human being, even one in desperate straits.[37]

## 1.4 Some Final Questions

### 1.4.1 Spiritual Acts Personism

It would be naïve of course to pretend that the resources of Catholic moral theology are united and available to counteract those who hold that people lacking the exercise of rational autonomy do not matter or those who deny that some people should receive assisted feeding. The Christian tradition boasts its own versions of libertarian-liberalism, utilitarian-liberalism and autonomy-personism. These include situation ethics, some fundamental option theories, proportionalism and even some more 'traditional' but in my view ultimately dualistic approaches. John Paul II sought to counter some of these approaches in *Veritatis Splendor, Evangelium Vitæ, Fides et Ratio* and his 2004 Allocution. I suspect they are now waning in the theologates much as their counterparts have been waning in the secular academy for decades. So I will not critique them here. Nonetheless it should be noted that theologians operating out of such approaches have long advised not feeding and hydrating those who lack the exercise of rational autonomy,[38] and some of them may well still be advising this.[39] Following the 2004 Allocution some theologians have flatly denied its teaching authority, calling the pope's position 'theologically erroneous', 'irresponsible', 'insulting' and 'mischief-making at the Vatican', questioning who really wrote it and whether the Pope was well enough to know what he was pronouncing, and recommending that people just ignore it (Reilly, 2005, nos. 34–35). The history of this kind of advice means that, ironically, some *Catholic* health, aged and palliative care providers may be *more* inclined to require rational autonomy (or the reasonable prospect of a return to exercising it) as a prerequisite for feeding than some of their secular counterparts.

A theological counterpart to the 'autonomy-personism' which can be found in several of these authors is what might be called 'spiritual-acts-personism.' This view holds that the human person is capable of many acts but that those directed towards securing the goods of the body (nourishment, exercise, healthcare . . .) are intrinsically inferior to, merely instrumental for and entirely ordered towards specifically human 'spiritual acts' (thought, contemplation, choice, worship, etc.). Some go further in saying that only someone who can now or will in the future have spiritual experiences or perform spiritual acts is fully a person or fully alive; others distinguish between personal-social-spiritual death and animal-biological death. According to these views, keeping someone alive when he can no longer have spiritual experiences or perform *spiritual acts* or at least engage in relationships with others is at best pointless, because life has lost its point. It may even be cruel or 'blasphemous' as we are depriving the person of 'release' and delaying their entry into heaven. This view of the purposes of human bodily life obviously has implications for a much broader range of people than those who suffer 'PVS.'

The view that those incapable of spiritual acts should no longer receive life-sustaining care has something in common with an ancient line of thought that runs thus: if heaven is so good, why not go there now? The answer to it given by Schopenhauer, among others, was that suicide is an assertion of self at its strongest, hardly the kind of gentle acceptance of God's will that will open up heaven to us; homicide mirrors this willfulness. Secondly, spiritual-acts-personism is often mind-body dualist and amenable to many of the same criticisms which I noted above have been levelled against more secular dualisms. Thirdly, it parallels the élitism of autonomy-personism which I have already critiqued in 1.2.1 and 1.2.2 above. Fourthly, even if sub-spiritual goods serve spiritual ones, it does not follow that they lack any value in themselves for human life may be good in itself and therefore worthy of protection and nurture, even when it fails to serve some higher 'spiritual' good such as prayer.[40] Those who continue to value and sustain such an impoverished instance of human life may do so not out of an irrational attachment to biological life but rather out of love and respect for the person whose life it is, even if that person cannot consciously experience that act of love (Grisez, 1996). Fifthly, a concern to preserve the norm against killing—including killing by neglect to provide effective and non-burdensome care—serves not only the person whose life is sustained but also the common good of the whole community, especially those most vulnerable and at risk of homicidal omissions of care (Barry, 1989). Sixthly, if heaven is the presumed end of the person with serious cognitive impairment—and that is quite a presumption—and if denying tube-feeding to such a person is a kindness, it is hard to see why spoon feeding should not also be stopped. Indeed, why not expedite heaven for them with more active measures? Seventhly, can it be presumed that those suffering from 'PVS' or like conditions have no 'spiritual experiences' or engage in no 'spiritual acts'—whatever these precisely mean? If John the Baptist leapt in Elizabeth's womb at the coming of the embryonic Jesus (*Lk* 1:44), spiritual experience may not be reserved to the rationally autonomous. Finally, can we assume that the 'suffering' of such people is purposeless for them and for others? Is no purgatory possible for people while on earth? Can the situations of such people not be a spiritual opportunity for others, to demonstrate reverence despite repugnance, justice and charity, compassion and care, which might contribute to their own good too?

### 1.4.2 But No-one Wants to Live That Way!

Still we might say: *but no one wants to live that way!* Of course not. As I argued above, there are countless awful situations which we would not want to be in or want others we loved to be in. No-one would want to suffer double incontinence or progressive dementia or persistent unresponsiveness. No reasonable person would wish such things on others. To live with such a condition in prospect or present for ourselves or others may well try our hope, courage, patience, perseverance, love. It may evoke in us repugnance, anxiety or fear. It may exhaust us physically, emotionally and spiritually. But all this is a very different matter to saying that our life (or theirs) would no longer be 'worth living' and that our death (or theirs) would be no loss; or that we (or they) would lose our 'dignity' or that our life (or theirs) would lose its 'sanctity'; or that others should then hasten our deaths or neglect to give us even basic care.

To put it another way: is the action of a Teresa of Calcutta or Catherine of Siena irrational, even cruel, if it lengthens 'the kind of life no-one would want for themselves or those they loved'? John Paul II in *Evangelium Vitæ* recognized that those who seek euthanasia may do so out of anguish, desperation or conditioning, thus lessening or removing their subjective responsibility and that those who engage in euthanasia may be motivated by pity rather than a selfish refusal to be burdened with the life of someone who is suffering (John Paul II, 1995, no. 15). He nonetheless argued that euthanasia is 'false mercy', indeed 'a disturbing perversion of mercy.'

> True 'compassion' leads to sharing another's pain; it does not kill the person whose suffering we cannot bear. Moreover, the act of euthanasia appears all the more perverse if it is carried out by those, like relatives, who are supposed to treat a family member with patience and love, or by those, such as doctors, who by virtue of their specific profession are supposed to care for the sick person even in the most painful terminal stages...The height of arbitrariness and injustice is reached when certain people, such as physicians or legislators, arrogate to themselves the power to decide who ought to live and who ought to die...Thus the life of the person who is weak is put into the hands of the one who is strong; in society the sense of justice is lost, and mutual trust, the basis of every authentic interpersonal relationship, is undermined at its root (John Paul II, 1995, no. 66).

This 'false pity' is contrasted with 'the way of love and true mercy' which recognizes that in the face of 'the supreme confrontation with suffering and death', when all are tempted 'to give up in utter desperation', what is really called for is 'companionship, sympathy and support in the time of trial...help to keep on hoping when all human hopes fail' (John Paul II, 1995, no. 66).

### *1.4.3 Dying or as-Good-as-Dead?*

It has long been recognized that when someone is imminently dying treatments aimed at prolonging life are no longer appropriate and some forms of care should be scaled down even as others might be increased. However, as I have argued elsewhere, to label 'PVS' and like patients as 'dying', 'as good as dead' or as having a 'lethal pathology' and to call withholding nutrition and hydration from them 'allowing a natural dying process to proceed' is often confused and inclines people to unethical behaviour (Fisher, 1997). No one denies that 'PVS' and like conditions are very serious ones but how can someone like Terri Schiavo who, had she been fed, would probably have lived for years be said to have been dying? Is 'dying' simply a tag we use for a special class of patients who will, once so labelled, be denied even fairly basic care and so die sooner rather than later? The tag 'dying patient' thus becomes a self-fulfilling prophecy or, more sinisterly, a death sentence.

Every human being suffers from the 'life-threatening condition' that if denied water and nourishment he or she will undergo a 'natural dying process.' Some people (diabetics, babies, the handicapped) are more dependent than others upon technology or other people's energies for the satisfaction of such basic needs. But people who need help to achieve feeding and hydration are not dying people unless we choose to make them so: they are alive like any other organism, with the same need for food and water.

## 1.4.4 Food and Death in Contemporary Culture

In this paper I have offered the beginnings of a metaphysics and theology of food and feeding. I would like to offer one additional, somewhat provocative thought: that the last civilization we should trust about feeding issues is probably our own. While millions starve we have an eating crisis in the West; childhood obesity, adult obesity and diabetes are at epidemic proportions; we are unable to sit at table together at home and yet become compulsive diners as soon as we go out the door; for all the obsession with 'health food' our supermarkets and takeaways maximize the unhealthy; we are subject to endless diet fads, stomach stapling and more sinister pressures to anorexia or bulimia; fat-reducing gym régimes and fat-extracting surgery are now major household expenditures; Cher and Michael Jackson and countless others have their bodies remade, some even into the appearance of the opposite sex; we have trouble fitting into one airplane seat while our 'models' starve themselves to death; binge drinking is a regular entertainment, especially for the young; alcohol abuse amongst adults breaks many bones, relationships and lives; and so we might go on.

Meanwhile the same consumer culture has a very strange relationship with death and dying.[41] There are countless signs of *denial* in this area: the desperate and ultimately futile attempts to delay or eliminate the signs of aging (again through cosmetic surgery, gyms, fantasies and similar techniques to those used to evade the implications of immoderate eating); attempts at cryopreservation, genetic enhancement and the like to obtain eternal mortal life; the relegation ('warehousing') of the frail elderly and dying to institutions where 'out of sight' is 'out of mind.' Yet the same death-denying culture is often a 'culture of death', using death as an instrument of the strong against the weak and a 'solution' to suffering of various kinds, killing incalculable numbers through warfare and preventable starvation, surgical abortion, abortifacient drugs, embryo exploitation, neglect of the disabled newborn, drug trafficking and substance abuse. Our consumer culture hopelessly seeks to 'tame' death into one more product when it can no longer be denied by precisely controlling its time and quality. My thought here is that a civilization that is so dysfunctional when it comes to eating and drinking, death and dying, should be extra careful about initiating new life and death policies about withholding food and drink from the frail and dying.[42]

## 1.4.5 What Else Should We do for Those Lacking the Exercise of Rational Autonomy?

In this paper I have not considered all the complexities of applying the principle that we should *ordinarily* feed patients, even 'PVS' and like patients, even if this requires some technical assistance such as a PEG-tube. There will be questions of whether a person may ethically volunteer in advance not to receive such help, and for what reasons and when; whether a surrogate may decide on behalf of the unconscious person to make such a 'sacrifice'; what should be done when health resources are limited and decisions must be made about who gets what; what rôle carers, family members, guardians, courts and the state should play in decisions about assisted feeding; and so on. I have been addressing a prior question: why we would even trouble ourselves about such matters if a patient can no longer exercise rational autonomy. Even if we resolve the why-people-matter issue and the assisted feeding dilemma, it will not be enough. There will be more to do for people who suffer 'PVS' or other major cognitive impairments. In the last part of his 2004 Allocution John Paul invites such thinking, pointing out:

> *We must promote positive action* as a stand against pressures to withdraw hydration and nutrition and so put an end to patients' lives. It is necessary, above all, to support the families of those suffering this terrible condition. They must not be left alone with their heavy human, psychological and financial burden. Although the care for these patients is not, in general, particularly costly, society must allot sufficient resources for this care. There must be appropriate, concrete initiatives such as: a network of 'awakening centres', with *specialized treatment and rehabilitation* programmes; *financial support and home assistance* for families; *facilities to accommodate* those who cannot be cared for at home; and *respite for families* at risk of psychological and moral burn-out ... *Spiritual counselling and pastoral aid* are particularly important as they help the family find meaning in this apparently hopeless situation (John Paul II, 2004, no. 6).

## Notes

[1] This is a modified version of "Why do those who cannot exercise rational autonomy matter?" a paper given to *Catholic Bioethics in the Public Forum*, a Symposium of the International Association of Catholic Bioethicists, Order of Malta and the John Paul II Institute for Marriage and the Family, Melbourne, 27 June 2005. My thanks to those whose feedback at the symposium contributed to the improvement of this chapter and to Brett Doyle for his editorial assistance.

[2] See: *Airedale NHS Trust* v *Bland.*, 1993; *Re D, Medical Law Review*, 1997; *Northridge* v *Central Sydney Area Health Service*, 2000, per O'Keefe J. For the string of legal opinions and judgment in the Schiavo case see: http://news.findlaw.com/legalnews/lit/schiavo/ Commenting upon some of these cases: Finnis, 1993; Fisher, 1993a, 1993b; Keown, 1993.

[3] See *Roe* v *Wade*,1973; *Webster* v. *Reproductive Health Services*,1989; *Planned Parenthood* v. *Casey* 1992.

[4] This anecdote is based on memory and class notes, but is not an exact transcript. My apologies for any inaccuracies.

[5] Quoted in Johnston, 2005.

[6] The first view would seem to ground the case for "voluntary euthanasia" only, but not, strictly speaking, euthanasia of the incompetent who have never expressed a wish for it. The second view,

on the other hand, would seem to allow (indeed require) some "involuntary euthanasia", even of patients who do not want to die, but whom we judge would be better off dead or whom we'd be better off with dead.

[7] See John Paul II, 1993, Ch. 2. This aspect of John Paul II's magisterium has already repeatedly been cited by Benedict XVI. See also John Paul II, 1995, nos. 19, 32ff, 61ff.

[8] John Paul II, 1995, no. 64; cf. nos. 14 & 15 on the tendency to evade suffering by killing the person who suffers; no. 23 on "the censorship of suffering" as something always to be avoided by whatever means necessary; no. 51 on the meaning Christ gives to suffering; and no. 67 on the way of love and true mercy.

[9] See also John Paul II, 1995, no. 43; Benedict, 2006.

[10] See Boyle, 1992; Andrew Fergusson, 1993; Grisez, 1996; May, 1991; McMahon, 2004.

[11] See Ratzinger, 1991; 1996a, 1996b.

[12] See e.g. Aquinas, *Summa theologiæ* Ia, 75 & 76 where he argues that men are not their souls or the operations of their souls (such as reason), but the unity that is their body and soul; and that even after death the separated soul is not the 'person' but awaits the restoration of its embodiment. In Ia IIæ 94, 2 he argues that the living/vegetative, sensitive/animal and rational/human aspects of the human person all make their moral demands and, because man is a rational living animal, these several requirements—including the requirement to preserve human life—are matters for practical reason. Aquinas was concerned to contest the views of the Cathars and Albigensians who argued that the soul was a prisoner of the body craving release and that all that mattered about the human person was his soul or mind; it was to counter this perennial dualism which Aquinas' Dominican Order had only recently been founded; his mentor Augustine had met and contested it in the form of Manicheism.

[13] See Wojtyla, 1981; John Paul II, 1997; John Paul II, 1993, nos. 46–50.

[14] See Midgley, 1979; Finnis, 1998; Middleton, 2005.

[15] John Paul II, 1995, no. 19 notes the contemporary tendency to exalt the concept of consciousness or subjectivity "to an extreme" and to recognize "as the subject of rights only the person who enjoys full or at least partial autonomy and who emerges from a state of total dependence on others."

[16] John Paul II, 1995, no. 12 suggests that "it is possible to speak in a certain sense of a *war of the powerful against the weak:* a life which would require greater acceptance, love and care is considered useless, or held to be an intolerable burden, and is therefore rejected in one way or another."

[17] See John Paul II, 1995, no. 66: "The choice of euthanasia becomes more serious when it takes the form of a *murder* committed by others on a person who has in no way requested it and who has never consented to it. The height of arbitrariness and injustice is reached when certain people, such as physicians or legislators, arrogate to themselves the power to decide who ought to live and who ought to die. Once again we find ourselves before the temptation of Eden: to become like God who "knows good and evil" (*cf. Gen* 3:5). God alone has the power over life and death: "It is I who bring both death and life" (*Dt* 32:39; cf. 2 *Kg* 5:7; 1 *Sam* 2:6). But he only exercises this power in accordance with a plan of wisdom and love. When man usurps this power, enslaved by a foolish and selfish way of thinking, he inevitably uses it for injustice and death. Thus the life of the person who is weak is put into the hands of the one who is strong; in society the sense of justice is lost, and mutual trust, the basis of every authentic interpersonal relationship, is undermined at its root."

[18] John Paul II, 1995, no. 11 notes that "the value of life can today undergo a kind of 'eclipse' ... as is evident in the tendency to disguise certain crimes against life in its early or final stages by using innocuous medical terms which distract attention from the fact".

[19] *Cf.* John Paul II, 1995, no. 66: "To concur with the intention of another person to commit suicide and to help in carrying it out through so-called 'assisted suicide' means to cooperate in, and at times to be the actual perpetrator of, an injustice which can never be excused, even if it is requested. In a remarkably relevant passage Saint Augustine writes that 'it is never licit to kill another: even if he should wish it, indeed if he request it because, hanging between life and death, he begs for help in

freeing the soul struggling against the bonds of the body and longing to be released; nor is it licit even when a sick person is no longer able to live."

[20] See John Paul II, in *Evangelium Vitæ* nos. 40–41 on the negative and positive implications of the inviolability norm and no. 57 which defines the negative moral absolute as a matter of Catholic dogma after rehearsing the philosophical and theological arguments for this.

[21] See Fisher, 1993c.

[22] *Cf.* Gratian, *Decretum,* C. 21, dist. LXXXVI (ed. Friedberg I, 302). According to Flannery this axiom is also found already in PL 54, 591 A (*cf.* in *Antonianum* 27 [1952] 349–366).

[23] *Cf.* Miles, "Nourishment and the ethics of lament," *Linacre Quarterly* 56(1) (Aug 1989), 64–69.

[24] See John Paul II, 2004, no. 4: "the administration of water and food, even when provided by artificial means, always represents a natural means of preserving life, not a medical act. Its use, furthermore, should be considered, in principle, ordinary and proportionate, and as such morally obligatory, insofar as and until it is seen to have attained its proper finality, which in the present case consists in providing nourishment to the patient and alleviation of his suffering."

[25] See John Paul II, 1995, nos. 34–36, 53–55.

[26] See also Australian Catholic Bishops and Catholic Health Australia, 2004.

[27] Bishop Sgreccia's statements—that the removal of tube-feeding from Terri Schiavo and like patients amounts to "direct euthanasia" and "a cruel way of killing someone" were reported on Vatican Radio on 11 March 2005 and Catholic New Service of the same day. Several American bishops agreed: see "Florida Bishops: The Care of Terri Shiavo," *Origins,* 34(39) (17 Mar 2005), 632; and surrounding editions of *Origins.*

[28] See e.g. *Deut* 10:18f; 14:29; 24:17ff; *Prov* 25:21; *Sir* 4:1–6; *Job* 22:7; *Isa* 58:6ff; *Ezek* 18:7,16; *Tobit* 1:16f & 4:16; *Mt* 10:42; 25:31ff; *Mk* 9:41; *Lk* Ch. 10; *Rom* 12:20; *Jam* 2:15ff; 1 *Jn* 3:17.

[29] See e.g. *Gen* 43:34; 1 *Kings* 17:4,8; 21:7; *Ps* 81:16; 107:9; 146:7; *Eccl* 8:15; *Ecclus* 31:30ff; *Ezek* 34; *Mt* 6:25ff *par; Lk* 15:23.

[30] *Cf. Dt* 21:18ff; *Prov* 21:17; 23:20f; 28:7; *Ecclus* 31:12ff; 37:32ff; *Tit* 1:12.

[31] *Cf.* John Paul II, 1995, no. 43.

[32] *Cf Lk* Chs. 10–14.

[33] *Cf. Mk* 1:41; 6:34 ff *par.*; 9:22; *Mt* 18:23 ff; 20:34; 24:45 ff; *Lk* 10:33 ff; 15:11 ff; 16:1 ff; *Jn* 21.

[34] *Cf. Jn* 13:35.

[35] In this essay I have focused principally on the Gospel as a source of a theology of food and feeding. A more complete treatment would survey other important Christian sources, such as the Old Testament, the Epistles, the works of the Fathers and the Scholastics, sacramental practice (especially the Rites of Care for the Sick, Dying and Recently Deceased), sacred art and music etc.

[36] *Cf.* 1 *Cor* Ch. 12; *Evangelium Vitæ* no. 67.

[37] John Paul II pointed out in *Evangelium Vitæ* no. 65 that depending on the circumstances euthanasia—by action or omission—"involves the malice proper to suicide or murder".

[38] E.g.—from many different starting points—Benedict Ashley, Edward Bayer, Thomas Bole, Philip Boyle, Daniel Callaghan, Robert Craig, Richard Devine, Eileen Flynn, Norman Ford, Kevin Kelly, Joseph Kukara, Daniel Maguire, James McCarthey, Richard McCormick, Thomas O'Donnell, Kevin O'Rourke, John Paris, Thomas Shannon, Andrew Varga, James Walter, Kevin Wildes and Anthony Zimmerman. See e.g. articles in Wildes et al. , 1992.

[39] In a recent article Patrick Reilly (2005) noted the continuing efforts of some in Catholic academies, especially some connected with Boston College, to promote laws and a culture in which those at a very low ebb are denied nutrition and hydration.

[40] Thus Augustine, *De doctrina Christiana* I, 23.26 and Aquinas, in *Summa Theologiæ* IIa IIæ 25.4, 25.5, 25.12 argue that there is a duty to love one's own life and one's own body, even if these may at times properly resign oneself to losing one's (bodily) life (= "sacrificing oneself") for the sake of God, neighbour or one's own eternal destiny.

[41] See the very interesting analysis of Hayden Ramsay (2005) in a series of articles on death in *New Blackfriars* (Jan to Sept 2005), and the sources cited therein.

[42] See John Paul II, 1995, no. 64: "In this context the temptation grows to have recourse to *euthanasia*, that is, *to take control of death and bring it about before its time*, 'gently' ending one's own life or the life of others. In reality, what might seem logical and humane, when looked at more closely is seen to be *senseless and inhumane*. Here we are faced with one of the more alarming symptoms of the 'culture of death', which is advancing above all in prosperous societies, marked by an attitude of excessive preoccupation with efficiency and which sees the growing number of elderly and disabled people as intolerable and too burdensome. These people are very often isolated by their families and by society, which are organized almost exclusively on the basis of criteria of productive efficiency, according to which a hopelessly impaired life no longer has any value."

# References

*Airedale NHS Trust* v *Bland*. (1993). AC 789.

Australian Catholic Bishops and Catholic Health Australia. (2004). *Briefing Note on the Obligation to provide Nutrition and Hydration.* Available on-line at: http://www.acbc.catholic.org.au/bc/docmoral/2004090316.htm.

Barry, R., O.P. (1989). Feeding the comatose and the common good in the Catholic tradition. *Thomist, 53*, 1–30.

Benedict XVI. (2006). *Deus caritas est: Encyclical on Divine Charity.* Vatican City: Libreria Editrice Vaticana.

Boyle, J. (1992). The American debate about artificial nutrition and hydration. In L. Gormally (Ed.), *The Dependent Elderly: Autonomy, Justice and Quality of Care.* Cambridge: Cambridge University Press, 28–46.

Castle, A. (1994). *Quotes and Anecdotes.* Mystic CT: Twenty-Third Publications.

Catholic Health Australia. (2001). *Code of Ethical Standards for Catholic Health and Aged Care Services in Australia*, Red Hill: Australia.

Dworkin, R. (1993). *Life's Dominion: An Argument about Abortion, Euthanasia, and Individual Freedom*, New York: Knopf.

Fergusson, A. (1993). Should tube-feeding be withdrawn in PVS? *Journal of the Christian Medical Fellowship.* April, 4–8.

Finnis, J. M. (1993). Bland: Crossing the Rubicon? *Law Quarterly Review, 109*, 329–337.

Finnis, J. M. (1998). *Aquinas: Moral, Political and Legal Theory.* Oxford: Oxford University Press.

Fisher, A., O.P. (1992). The incarnation and the fully human life. *New Blackfriars, 73*, 396–407.

Fisher, A., O.P. (1993a). Old law and new ethics: Bland's Case and not feeding the comatose. *Law & Justice, 116/117*, 4–18.

Fisher, A., O.P. (1993b). On not starving the unconscious. *New Blackfriars, 74*, 130–145.

Fisher, A., O.P. (1993c). *Killing and Letting Die: What's the Difference?* London: Signum.

Fisher, A., O.P. (1995). *Jesus: Glutton and Drunkard.* Manchester: Blackfriars Publications.

Fisher, A., O.P. (1997). Should we starve the unconscious? *Australasian Catholic Record, 74*, 315–329.

Fisher, A. O.P. (2005). The ethics of care for those with post-coma unresponsiveness and related conditions. *Bioethics Outlook, 16*, 1–5.

George, R. (1993). *Making Men Moral: Civil Liberties and Public Morality.* Oxford: Oxford University Press.

Grisez, G. (1996). May a husband end all care of his permanently unconscious wife? *Linacre Quarterly, 63*, 41–46.

Grubb, A. & Kennedy, I. (Eds.) (1997). Re D, *Medical Law Review, 5*, 225–26.

Hauerwas, S. (1986). *Suffering Presence: Theological Reflections on Medicine, the Mentally Handicapped and the Church.* Notre Dame IN: University of Notre Dame Press.

John Paul II. (1993). *Veritatis Splendor: Encyclical on Certain Fundamental Questions of the Church's Moral Teaching.* Vatican City: Libreria Editrice Vaticana.

John Paul II. (1995). *Evangelium Vitæ: Encyclical on the Value and Inviolability of Human Life.* Vatican City: Libreria Editrice Vaticana.

John Paul II, (1997).*The Theology of the Body: Human Love in the Divine Plan.* Boston: Pauline Books.

John Paul II. (1998). *Ad limina* address of the Holy Father to US Bishops of California, Nevada and Hawaii. Available online at: *www.wf-f.org/JPII-Bishops-Life-Issues.html.*

John Paul II. (2004). *Allocution* to the participants to the international congress life-sustaining treatments and the vegetative state: Scientific advances and ethical dilemmas. Available on-line at: http://www.vatican.va/holy_father/john_paul_ii/speeches/2004/march/documents/hf_jp-ii_spe_20040320_congress-fiamc_it.html.

Johnston, P. (2005). Is it "murder" to pull Terri's feeding tube? *The Intellectual Conservative 22;* available on-line at http://www.intellectualconservative.com/article4226.html.

Keown, J. (1993). Courting euthanasia? Tony Bland and the law lords.' *Ethics & Medicine, 9,* 34–39.

May, W. (1991). Criteria for withholding or withdrawing treatment. *Linacre Quarterly, 57,* 81–90.

MacIntyre, A. (1984). *After Virtue* 2nd ed. Notre Dame: University of Notre Dame Press.

MacIntyre, A. (1999). *Dependent Rational Animals: why human beings need the virtues.* Chicago: Open Court.

McMahon, K. (2004). Catholic moral teaching, medically assisted nutrition and hydration and the vegetative state.*NeuroRehabilitation, 19,* 373–80.

Middleton, N. (2005). Aquinas, the enlightenment and Darwin. *New Blackfriars, 86,* 437–449.

Midgley, M. (1979). *Beast and Man: The Roots of Human Nature.* London: Methuen.

*Northridge* v *Central Sydney Area Health Service.* (2000). NSWSC 1241.

Miles, S. (1989). Nourishment and the ethics of lament. *Linacre Quarterly, 56,* 64–69.

National Conference of Catholic Bishops. (1994). *Ethical and Religious Directives for Catholic Health Care Services.* Washington DC: Bishops Committee on Doctrine.

*Planned Parenthood* v. *Casey.* (1992). 505 U.S. 833.

Pontifical Council for Health Care Workers. (1994). *Charter for Health Care Workers.* Available on-line at: http://www.lifeissues.net/writers/doc/hc/health_care_1.html.

Ramsay, H. (2005). In a series of articles on death in *New Blackfriars, 86,* 1001–1005.

Ratzinger, J. (1991). The problem of threats to human life. *L'Osservatore Romano, 8,* 2–4.

Ratzinger, J. (1996a). Truth and freedom. *Communio, 23,* 15–35.

Ratzinger, J. (1996b). The current state of faith and philosophy. *L'Osservatore Romano, 6,* 7.

Raymond of Capua. (1960). *Life of St Catherine of Siena.* Lamb, G. (Trans.). New York: P J Kennedy.

Reilly, P. (2005). Teaching euthanasia. *Crisis Magazine, 23,* 28–35.

*Roe* v *Wade.* (1973). 410 U.S. 113.

Rorty, R. (1991). *Essays on Heidegger and Others.* Cambridge: Cambridge University Press.

Stark, R. (1996). *The Rise of Christianity: A Sociologist Reconsiders History.* Princeton: Princeton University Press.

Tonti-Filippini, N., Fleming, J., & Walsh, M. (2004). Twenty propositions. *Human Life Review 30,* 705–712.

US Bishops' Pro-Life Committee (1992). Nutrition and hydration: moral and pastoral reflections. *Origins, 21,* 705–712.

Vatican Council II (1965). *Gaudium et Spes.* Vatican City: Libreria Editrice Vaticana.

*Webster* v. *Reproductive Health Services.* (1989). 491 U.S. 397.

Wildes, K. Harvey, J.C., & Abel, F. (Eds.). (1992). *Birth, Suffering and Death: Catholic Perspective at the Edges of Life.* Dordrecht: Kluwer.

Wojtyla, K. (1981). *Love and Responsibility.* New York: Farrar, Straus & Giroux.

# Chapter 2
# Are We Morally Obliged to Feed PVS Patients Till Natural Death?

Michael Degnan

On March 20th, 2004, Pope John Paul II rocked the world of Catholic biomedical ethics in an address to the participants of the International Congress on "Life-Sustaining Treatments and Vegetative State: Scientific Advances and Ethical Dilemmas," when he delivered the first explicit papal statement affirming the obligation to provide food and water for patients diagnosed as being in a vegetative state (vs) persistently or permanently (John Paul II, 2004). The Pope reasoned that since food and water, even when administered artificially, are not medical acts, but natural means of sustaining human life, the practice of providing food and water to vs patients was in principle ordinary and proportionate, and therefore in principle, morally obligatory. If in a specific case artificially delivered nutrition and hydration (ANH) provides neither nourishment nor any pain reduction, then ANH fails to offer a reasonable hope of benefit and is not morally obligatory.

The Pope is less clear about what conditions make providing such care disproportionate. Unlike the 1992 US Bishops' Pro Life Committee resource paper, "Nutrition and Hydration: Moral and Pastoral Reflections," the Pope did not explicitly state that excessive burdens on the patient's family and community could justify withdrawal of feeding provided that the aim was not the death of the patient (Pro Life Committee, 1992).[1] He was not blind to these burdens, for he recognized that the psychological and financial burdens on families providing such care for years, can be so great that communities are morally obliged to help families carry this burden. He explicitly affirms that the decreasing probability of recovery after one year in the vegetative state "cannot ethically justify the cessation or interruption of *minimal care* for the patient, including nutrition and hydration" (John Paul II, 2004, no. 4). He explicitly denies that considerations of quality of life can take precedence over general principles about providing ordinary care, for these concerns are the results of psychological, social or economic pressures. The ground for providing this ordinary care is "*A man, even if seriously ill or disabled in the exercise of his highest functions, is and always will be a man*, and he will never become a 'vegetable'

M. Degnan University of St Thomas,
St. Paul, MN
Email: mjdegnan@stthomas.edu

C. Tollefsen (ed.), *Artificial Nutrition and Hydration:*
*The New Catholic Debate*, 39–60. © Springer 2008

or an 'animal'" (John Paul II, 2004, no. 3). In light of his acknowledgement of the community's obligation to help the vs patient's family and his recognition of the psychological, social and economic pressures, it is reasonable to interpret John Paul II as believing that with community support the burden of supplying ANH is not excessive and is morally obligatory when it provides nourishment or pain reduction.[2] The Pope's address sparked criticism from a wide range of Catholic bio-ethicists.[3]

In this paper I offer a philosophical defense of John Paul II's teaching that the human community is morally obliged to provide and the patient morally obliged to accept, food and water when a patient for whom death is not imminent is diagnosed as being in vegetative state of any duration, provided that the patient can assimilate nourishment and water without compromising other necessary life functions. Central to my argument is the special relationship between the community and its members' need for food and water. Alasdair McIntyre has argued in *Dependent Rational Animals (DRA),* that the natural dependence and vulnerability of human beings requires social practices of giving and receiving where uncalculated giving to dependent humans is a necessary condition for developing a community of independent practical reasoners (MacIntyre, 1999). In this paper I contribute to the deliberation among independent practical reasoners about what the common good requires of us as individuals and as a community in providing food and water to vs patients. In *DRA* MacIntyre does not endorse this conclusion, nor does what he write entail such an endorsement. His theory of the common good and the role of the virtues of acknowledged dependence in forming a community of independent practical reasoners who live humanly flourishing lives can be used to make a case for the community's obligation to construct and provide a network of giving and receiving relationships that under certain conditions require providing food and water to vs patients until their natural death. I will argue that the community practice of providing ANH to vs patients contributes to the patients' human flourishing as well as the community's, and that the community owes such care to these patients. The absence of a practice of giving food and water to vs patients disrupts the relationship between the individual's good and the common good. This understanding of the common good implies changes in our social practices so that families and friends of such brain-injured patients are supported by the community. I will sketch two models of how social practices can change to systematically support vs patients and their families in providing this care.

In addition to this defense, I respond to criticisms that find the Pope's teaching mistaken. One criticism maintains that the mere preservation of physical life is not sufficient to obligate someone to use specific means to sustain life, since the moral imperative to help the individual holds only if the individual has a potential for cognitive-affective function. On this view the duty to preserve life is conditioned on what good is to be achieved by continuing to live. A second criticism recognizes that sustaining life is a benefit for the vs patient, but that this benefit can be outweighed by excessive economic and psychological burdens borne by the patient or the patient's family or community.

## 2.1  The Argument

The Pope's central argument is stated in paragraph 4 of his allocution.

> I should like particularly to underline how the administration of water and food, even when provided by artificial means, always represents a *natural means* of preserving life, not a *medical act.* Its use, furthermore, should be considered, in principle, *ordinary* and *proportionate*, and as such morally obligatory, insofar as and until it is seen to have attained its proper finality, which in the present case consists in providing nourishment to the patient and alleviation of his suffering. The obligation to provide the "normal care due to the sick in such cases" (Congregation for the Doctrine of the Faith, *Iura et Bona*, p. IV) includes, in fact, the use of nutrition and hydration (cf. Pontifical Council "Cor Unum," *Dans le Cadre*, 2, 4, 4; Pontifical Council for Pastoral Assistance to Health Care Workers, *Charter of Health Care Workers*, n. 120).

The Pope's argument can be expressed as a categorical syllogism:

1. All patient care that is ordinary and proportionate is morally obligatory for the patient to receive and the community to provide (obligatory care principle).
2. Artificial nutrition and hydration (ANH) for any patient in vs who can be nourished or relieved of suffering is ordinary and proportionate care.
3. So, ANH for any patient in vs who can be nourished or relieved of suffering is morally obligatory for the patient to receive and the community to provide.

Before considering the meaning and truth of the premises in the Pope's argument I want to narrow the subject class of vs patients for whom I claim that ANH is ordinary and proportionate care to vs patients for whom death is not imminent and for whom assimilation of food and water is possible without significant harm to bodily systems necessary for life. When death is imminent or when the feeding cannot be processed as nourishment or when the feeding threatens the life of the patient by causing life-threatening respiratory problems, for example, then the feeding is not beneficial to the patient and so is not ordinary care. The revised second premise states that ANH for all vs patients for whom death is not imminent and for whom assimilation of food and water is possible without significant harm to bodily systems necessary for life is ordinary and proportionate care. The conclusion's subject is modified in the same way.

## 2.2  Meaning of Terms

I begin by considering the meaning of key terms in the premises. Ordinary care offers a patient a reasonable hope of benefit. In many contexts this is distinguished from extraordinary treatments that do not offer a reasonable hope of benefit. In this context the phrase signifies basic care any sick person is owed as distinguished from medical treatments, since at this time there are no treatments for vs. Food, water, clothing and shelter typically comprise this basic care. A care is proportionate if the burdens imposed by the care on the patient, family and community are balanced by the benefit received by patient, family and community. ANH is food and water delivered to a patient through a tube that opens into the nose or one that opens directly into the stomach. In short, ANH is medically assisted supply of food and water (McMahon, 2004, p. 375).

The meaning of the phrase, "patients with vs" is more problematic than most ethical discussions recognize.[4] The definition that has most informed clinical decisions and bioethical discussion is that given in the 1994 consensus statement of the Multi-Society Task Force on PVS (MSTF):

> The vegetative state is a clinical condition of complete unawareness of the self and the environment, accompanied by sleep-wake cycles, with either complete or partial preservation of hypothalamic and brain-stem autonomic functions. In addition, patients in a vegetative state show no evidence of sustained, reproducible, purposeful, or voluntary behavioral responses to visual, auditory, tactile, or noxious stimuli; show no evidence of language comprehension or expression; have bowel and bladder incontinence; and have variably preserved cranial-nerve and spinal reflexes (1994, p. 1499).[5]

The first sentence defines vs from the first person perspective of the patient, no awareness of environment or self, along with two clinically observable behaviors, the sleep-wake cycle and autonomic functions of the hypothalamus and brain stem. Earlier in the document consciousness is defined as awareness of self and environment, hence this statement defines vs as an unconscious state. The second sentence lists additional clinically observable absences of voluntary behaviors combined with two kinds of reflexes and bowel and bladder incontinence. The clinically observable behaviors in the first sentence distinguish this state from patients in a coma who never are awake and whose autonomic functions are not completely operative. The second sentence is not strictly part of the definition, for it describes clinically observable behaviors or absences that are entailed by the lack of awareness and the functioning brain stem and hypothalamus. The lack of awareness of self and environment precludes voluntary action and language comprehension. The autonomic functions entail the reflex responses. The first sentence defines the condition, cites distinguishing clinically observable features and offers causally explanatory absences that account for the absences and behaviors described in the second sentence. Logically, the definition is first-rate.

Empirically, the definition is seriously flawed, however. The task force's denial of pain in vs patients is empirically unsupported. Dr. Alan Shewmon, Professor of Neurology and Pediatrics at the David Geffen School of Medicine at UCLA, reports that pain physiologists find that pain receptors end in the thalamus which is typically unharmed in pvs patients. He writes,

> All treatises on the neurophysiology of pain traced the anatomical pathway from the cutaneous nociceptors centrally, invariably ending not at the cortex but at the thalamus. Patients with strokes involving somato-sensory cortex lose tactile discrimination and joint position sense, but not the capacity to perceive and to localize pain (Shewmon, 1997, pp. 59–60).

Thalamic injury, however, can cause a distressing form of central pain. In the pain literature it is clear that the cortex's role in pain perception is merely modulatory and that the experience is mediated sub-cortically, but in the PVS literature these well known phenomena are systematically ignored. PVS patients often grimace to noxious stimuli and manifest primitive withdrawal responses. Advocates of the cortical theory write off such behaviors as mere brain-stem or spinal reflexes, but that dismissive attitude is based more on an *a priori* assumption than a scientific conclusion.[6]

The Task Force reports the grimaces and withdrawal behaviors interpreting them as brain-stem or spinal reflexes, not as indications of pain. Given that the thalamus and sub-cortical areas are not typically implicated in vs brain injury, these behaviors count as evidence of thalamic induced pain mediated by intact sub-cortical areas. If the pain physiologists are correct, vs patients exhibiting such behaviors very likely have pain experiences.

For those who believe cortical activity is necessary for pain, recent work has shown evidence of cortical activity in patients in vs longer than one year when exposed to noxious stimuli. In 2002 S. Laureys' research group published evidence that noxious stimuli activated midbrain, contra-lateral thalamus, and primary somatosensory cortex in every patient studied that met the clinically observable criteria for vs (Shewmon, 2004a, p. 345; Laureys et al., 2002). These researchers observed that the activated primary somatosensory cortex was functionally disconnected from the inactivated secondary somatosensory and limbic cortices (Kassubek et al., 2003).

Kassubek's research group discovered similar pain-induced activation of primary sensorimotor cortex in post-anoxic patients as well as activation of secondary somatosensory cortex, insula, cingulate and other association areas (Kassubek et al., 2003). This research group concluded, "the regional activity found at the cortical level indicates that a residual pain-related cerebral network remains active in long-term VS patients."[7] The case for vs patients having the capacity for pain experience is significant.

By contrast, the MSTF's evidence for unawareness of the environment and self is weak. The MSTF offers three lines of evidence for its conclusion.

> First, the motor or eye movements and facial expressions in response to various stimuli occur in stereotyped patterns that indicate reflexive responses integrated at deep subcortical levels rather than learned voluntary acts. Second, positron-emission tomography of pvs patients reveal regional cerebral glucose metabolism levels lower than patients who are aware or in a locked-in state. These metabolism levels are as low as those in patients during deep general anesthesia. Third, all available neuropathological exams of the brains of patients with observable features given in the definition "show lesions so severe and diffuse that awareness would have been highly improbable, given our biologic understanding of how the anatomy and physiology of the brain contribute to consciousness" (MSTF, 1994, p. 1502).

The first evidential line is consistent with unexpressed awareness of the environment. The MSTF gives no anatomical evidence for the claim that no cortical structures are involved in responding to sensory stimuli. The MSTF's judgment is an inference supported by the absence of voluntary behavioral responses.[8] It is possible that the patient may competently register the stimuli, but lack sufficient coordination between functioning cortical areas to muster a voluntary response to the stimuli.[9]

Recent neurological study supports this possibility. Using the distinction between competence and performance from linguistic theory Schoenle and Witzkey discovered an event-related potential in the non-injured brains of humans correlated with grasping the meaning of anomalous semantic sentences like "The coffee is too hot to fly." They found that same event-related potential in some vs patients (Schoenle & Witzkey, 2004, p. 341). They examined 120 brain-damaged patients who belonged to one of three classes for less than one year: vs, near-vs and not-vs. The vs individuals exhibited the clinical criteria of its class. The near-vs group showed at least one of the following: habituation, eye fixation, or visual pursuit. The brain-injured not-vs-group responded to commands and could initiate purposeful actions. When patients were presented with anomalous semantic sentences, 12% of vs patients and 76% of the near-vs patients and 91% of not-vs patients responded with the trademark event-related potential N400. The researchers interpreted this as evidence of a strong likelihood that some vs and most near-vs patients had linguistic competence for understanding, but were unable to process that understanding into a communicative expression. Here is concrete evidence that sufficient cortical structure may remain in vs and near-vs patients for awareness of environment without ability to communicate such awareness.

The MSTF's second line of evidence for denying awareness of environment for the vs patient is also inadequate, for the consciousness-and pain-suppressing effects of anesthesia are more likely due to its depressing effects on the brain stem, rather than the cortical area, (Shewmon, 2004a). Since the brain stem is functional in vs patients, the comparison with anesthetized patients is weak. The fact that glucose metabolisms are similar between anesthetized patients and those in vs is not proof of lack of consciousness. First, some vs patients have glucose levels as high as 65 % which are comparable to the levels of sleeping persons, not anesthetized ones (Shewmon, 2004a, and Shewmon, 2004b, p. 345; see also Beuthien-Baumann, Handrick, & Schmidt, 2003). Second, some vs patients who have recovered consciousness have done so with glucose levels as low as anesthetized patients (Laureys et al., 2000).

The third line of evidence for the denial of awareness in vs patients was the report's study of the medical literature of the twenty years prior to 1994 That biological understanding has changed dramatically in the thirteen years since MSTF's publication as the results of the studies cited above show. I have already shown that contrary to MSTF, cortical structures associated with pain awareness are activated in vs patients, including those lasting longer than one year. Another research team found three of five patients meeting the clinically defined features of vs to have residual cerebral activity correlated with sensory responses to stimuli (Schiff et al., 2002). These cortical activations were correlated with isolated behavioral patterns and metabolic activity. According to Schiff, these patterns of preserved metabolic activity constitute new evidence of the modular nature of individual functional networks that underlie conscious brain function. Finally, while activation of these secondary cortices have been thought to be correlated with conscious perception, Shewmon's research group found three girls with congenital loss of cortical structure along with intact and functioning brain stem who "interacted adaptively with the environment in a variety of ways, indicating beyond doubt the presence of some form of subjective consciousness," with no activation of any sensory or limbic cortex (Shewmon, 2004b, p. 345; Shewmon, Holmes, & Byrne, 1999). Most vs patients are not congenital, so Shewmon's work must be confirmed in acute post-traumatic vs patients. Still, Shewmon's study falsifies the claim that cortical activation is necessary for awareness of environment and self.

These studies combined with the experience of pain physiologists make a strong case for ascribing pain experience to vs patients who groan, grimace and exhibit withdrawal behavior from stimuli. This counts as awareness of environment. Hence, the MSTF is mistaken to define vs as unawareness of the environment. Given that we lack empirical correlates for awareness of self, the MSTF's definition of vs must be abandoned. These studies undermine the task force's assumption that all vs patients lack sufficient cortical organization to sustain such awareness. Since that assumption is the warrant for the task force's judgment that all vs patients are unaware of the environment and self, these more recent studies imply that there is little, if any, warrant for defining vs as unawareness of environment and self.

This review of relevant medical literature suggests that a more accurate definition of the vs state be limited to observable criteria. Patients in vs designate brain-injured patients who exhibit intermittent wakefulness evidenced by eye-openings or sleep-wake cycles on EEG, failure to respond to auditory, visual, tactile or noxious stimulation with any purposeful or voluntary behavior, absence of external behavior indicating language comprehension or expression, and only generalized physiological response to pain, abnormal posturing, deep breathing or significant perspiration, cranial nerve stimulation, along with bowel and bladder incontinence (American Congress of Rehabilitation Medicine, 1995; Giacino, 2004, p. 294; MSTF, p. 1500). Along with this definition, the medical community needs to inform the wider community that recent work in neurology and pain physiology give good reason to believe that vs patients' pain response behavior very likely indicates pain experience. This evidence-based interpretation will affect our judgments about our obligations to care for such patients. It is also worth nothing that given this definition, as well as the MSTF's, the near-vs patients are regularly classified as vs. The near-vs patients do not exhibit enough behavior to count as voluntary response to stimuli required for the minimally conscious state (MCS) (Giacino, Zasker, et al., 1997; Giacino, Ashwals, Childs, et al., 2002; Giacino, 2004, pp. 296–297).

According to the MSTF the vs state lasts but one month, after which the diagnosis changes to persistent vegetative state. Within a year of publication of MSTF, professional societies representing those in the rehabilitative field established a working group which published new conventions concerning these brain-injured patients (Giacino, 2004). These professionals observed that after one month in vs two different studies had shown that 50% of the patients would recover ability to voluntarily respond to stimuli within one year of injury. Thus they recommended eliminating persistent vegetative state as a diagnosis. They also instituted the practice of describing a vs patient with cause of injury and length of time post injury, e.g. post-traumatic vs 8 months. These societies affixed the designation, "permanent vegetative state" to post-traumatic vs patients who had not recovered after one year since the statistical likelihood of recovery at that point was low. Those injured in non-traumatic states were deemed permanent at earlier times. Since this phrase is a probable prognosis rather than a diagnosis, I will not use the phrase, "permanent vegetative state".[10] I will use the phrase, vegetative state, (vs) to refer to the brain injured state designated by the observable clinical criteria.[11] I will not use the terms persistent or permanent vegetative state unless quoting others.[12] Finally, I will assume that ANH is ordinary and proportionate for the first 12 months post vs diagnosis of those for whom death is not imminent and for whom assimilation of food and water is possible without significant harm to bodily systems necessary for survival. The ground for that judgment is the significant likelihood of recovery of voluntary response during that time period.

## 2.3 Support for Premise 1

I consider support for the premises of the Pope's argument. Ordinary and proportionate care for patients is obligatory for a community to provide because the community's commitment to seeking the good of each of its members requires it. Providing ordinary and proportionate care to patients is by definition care that has a reasonable hope of benefit for the patient and is worth the burdens incurred. The benefit of care is not the same as the benefit of treatment. Keeping a patient warm, clean, comfortable and fed does not treat or cure a patient's illness. Care practices of feeding, bathing, clothing and exercise benefit any living patient who can receive nourishment and engage in passive range of motion. Since these benefits are also described as proportionate, an intentional decision of the community not to provide such care is an intentional failure to will that patient's good. The community that does not establish practices of ordinary and proportionate care for its sick members is neither willing the good of, nor the good for, each of its members.[13] Such a community fails to legitimate its authority over its members. The moral ground for the community establishing binding practices on its members is the community's commitment to willing the common good, which includes willing the good of and the good for each of its members. Consequently, no such community can legitimately exercise authority over its members without accepting the obligation to establish practices of providing ordinary and proportionate care to its sick members.

Such care is obligatory for a patient to accept as a consequence of the patient being part of the community which wills the good for each of its members, including him or herself. This universal willing assures each member that the community will support her or him during conditions of dependency that can afflict any member at any time.

A more illuminating account of a community's obligation to its sick, weak and disabled members can be found in Alasdair MacIntyre's account of the virtue of acknowledged dependence which he believes is necessary for any community that aims at human flourishing. In *Dependent Rational Animals* MacIntyre's starting point is that necessarily, human identity is animal identity. Being a bodily organism is essential to being human. As animal organisms, human vulnerability and disability touch all humans at some time in their lives affecting their ability to flourish as humans. Thus, vulnerability and disability must enter into the account of what is good for humans. From this MacIntyre concludes that necessarily any account of human good includes virtues that enable humans to acknowledge the nature and extent of human dependence on others and to act in response to such dependence. Those are the virtues of acknowledged dependence. Communal practices of caring for the ill, old and disabled are not just responses to special interests; they constitute what it is to be a human being in a human community. Writes MacIntyre,

> What I am trying to envisage then is a form of political society in which it is taken for granted that disability and dependence on others are something that all of us experience at certain times in our lives and this to unpredictable degrees, and that consequently our interest in how the needs of the disabled are adequately voiced and met is not a special interest, ....but, rather the interest of the whole political society, an interest that is integral to their conception of their common good (MacIntyre, 1999, p. 130).

On MacIntyre's account the human good, a humanly flourishing community, is a common good, one that each member has a claim to share in to some extent and one that requires the involvement of the community. It is a community of humans aimed at flourishing interdependently, a community whose members seek to live with each other supporting and exercising characteristically human activities with their appropriate excellence. In order to achieve this good, human beings need virtues that enable them to function as independent and accountable practical reasoners and virtues that enable them to acknowledge the nature and extent of their dependence on others and to act in response to such dependence. This second kind of virtue is necessary for adult members to exercise toward their children so that the children can develop the virtues of independent practical reasoners. The social and political forms required for the achievement of this common good are roles and social practices which involve commitments that are in some respect uncalculating (parents to children and children to elderly parents, community to orphans, community to disabled) not only to a range of goods, but to particular others together with whom humans attempt to achieve the good. This commitment to others is doubly motivated: others are necessary to achieve the goal and the achievement is better when shared. The virtue of acknowledged dependence embodies this uncalculating generosity which is owed to particular community members. MacIntyre dubs it "just generosity." He writes, "Because I owe it, to fail to exhibit it, is to fail in respect of justice; because what I owe is uncalculated giving, to fail to exhibit it is to fail in respect of generosity" (MacIntyre, 1999, p. 120).

Practices of rendering ordinary and proportionate care to the sick are practices that are a constitutive part of the virtue of acknowledged dependence. This basic care that is proportionally beneficial is owed to members' in multiple ways. Nearly every sick member will have some blood relative who will have a special obligation to render care as brother, sister, parent or cousin. In addition, simply by being a member of the community one has a claim on a share in the life of the community and that claim requires that when sick, others in the community help restore that sick member to as full a sharing in community life as is possible. MacIntyre is eloquent in speaking of the value of helping the physically and cognitively disabled see opportunities for themselves which the community may have systematically excluded them from. Supporting the sick with basic care requires generosity of time, attention, financial and capital resources. Just as the care parents may be required to give to their children may be far more in time, money and energy than the parent received from her or his parents, so the ordinary care given to the sick may be more than has been given to the caregivers. There is no strict proportionality of giving and receiving. While the giving is uncalculated, MacIntyre warns that it must not be indiscriminate, for it must be to people who are really in urgent need so that others who really are in urgent need are not overlooked (MacIntyre, 1999, p. 126). With this background, I present my argument for premise 1.

> 4. Any practice that is a constitutive part of the virtue of acknowledged dependence is a practice necessary for a humanly flourishing community.
> 5. The community practice of providing ordinary and proportionate care to the sick is a constitutive part of the virtue of acknowledged dependence.
> 6. So the community practice of providing ordinary and proportionate care to the sick is a practice necessary for a humanly flourishing community.
> 7. Any community practice necessary for a humanly flourishing community is a morally obligatory practice.
> 8. So the community practice of providing ordinary and proportionate care to the sick is a morally obligatory practice (equivalent to premise 1).

MacIntyre has made a strong case for the necessity of the virtue of acknowledged dependence in fostering the excellent exercise of independent practical reasoning for our children's development to adulthood. Community practices of educating the young, teaching arts, crafts, sciences and the humanities, feeding, clothing and sheltering the poor are practices of the virtue of acknowledged dependence that are necessary for a humanly flourishing community. The exercise of the virtue of acknowledged dependence is an exercise of just generosity, a virtue that constitutes a flourishing human life. So this virtue is necessary both as an instrumental means for initiating and sustaining others as community members and as a constitutive means of living a humanly flourishing life. The practice of ordinary and proportionate care for the sick and disabled is necessary to assure members that their claim to participation in the community life is not diminished by what happens to them through disease, natural disaster, genetic mutation or injury. The existence of these practices is a necessary instrumental condition for many members' pursuit of their human flourishing. Without these practices the sick and disabled would be cut off from community life. Initiating, sustaining and participating in such practices constitute exercising the virtue of acknowledged dependence. Thus, premises 4 and 5 are strongly supported.

On this account the goal of a good human community, the common good, is being a humanly flourishing community. Therefore, any community practice necessary to achieve a humanly flourishing community is morally obligatory, provided conditions for triage resourcing do not obtain (in such cases some necessary practices could not in fact be exercised). Since a humanly flourishing community is a common sharing in community life, it is a good that all members have a claim to share in and that claim is not diminished by what happens to the member.[14] As a consequence, any practice that forfeited a member's claim to participation in that life for no reason other than the improvement of the rest of the community members would be prohibited. So, there are good reasons for believing premise 7 is true.

I have offered two accounts of the common good that support this first premise: On one account, if these cares are not morally obligatory, then the community fails to aim at the good of each of its members and so loses its moral legitimacy to exercise authority over all its members. On the other account, if these cares are not obligatory, the conditions required for humans realizing their common good are not met.

## 2.4 Support for Premise 2

Support for the second premise, the claim that ANH for patients in vs for whom death is not imminent and for whom assimilation of food and water is possible without significant harm to bodily systems necessary for life is ordinary and proportionate care rests in part on the judgment that ANH offers reasonable hope of benefit for the vs patient. Food and water to vs patients who can assimilate the food without otherwise being harmed are assured the benefit of continued life as a human being. In addition, ANH for vs patients must be understood as basic care, not medical treatment. While the insertion of a feeding tube through the nose or the stomach is a medical act, the subsequent delivery of food and water through a tube is not. It is simply giving food and water. Giving food and water to a sick patient who can be nourished or can be relieved of pain is ordinary care. This judgment was once assumed by professional medical societies, as a 1973 Report of the Board of Science and Education of the British Medical Research Council (BMRC) reveals. That report held that "The care involved in feeding is not, in a strict sense, medical treatment, even if provided in a hospital. It is ordinary care and is not given for any life-threatening disease" (BMRC, 1973). In 1986 the American Medical Association determined that ANH was treatment rather than care, and as such, that it could be withdrawn or refused at the request of the patient or the patient's family. Dr. Keith Andrews, Royal Hospital for Neuro-Disability and Institute for Complex Neuro-Disabilities, both in London, critiques this practice of calling feeding and hydrating patients a medical treatment:

> If tube feeding is treatment, what is being treated? Surely not the [patient's] brain damage. The food is not being given to correct any abnormal biochemical or pathological process, but to provide nutrition to normal tissues. To my mind the tube is simply a tool for daily living, similar to the specially adapted spoons that enable arthritic patients to feed themselves. The relevance of this is that in identifying tube-feeding as treatment we have found a convenient method of shortening the life of a disabled person (Andrews, 1992).

Food and water are very effective at nourishing the tissue and bringing the patient into homeostasis. Dr. G. Gigli, Department of Neurosciences, Ospedale "Santa Maria della Misericordia", Udine, Italy and Dr. M. Valente, Neurorehabilitation Unit, DPSMC, University of Udine, Udine, Italy, observe that ANH is well-accepted for prolonged treatments in cases of pharyngeal and esophageal stenosis, of amyotrophic lateral sclerosis (Lou Gehrig's disease) or of prolonged post-traumatic coma (Gigli and Valente, 2004, p. 317). If ANH is ordinary care for patients with these debilitating diseases, ANH is ordinary care for the class of vs patients for whom death is not imminent and for whom the nutritive effects of food and water can be had without significant harm to bodily systems necessary for life. This is a matter of fairness.

The vs patient, even if she lacks awareness of environment as well as self, receives benefit from ANH beyond the good of the patient's "vegetative operations," of digestion, respiration, and circulation of nutrient materials. Unlike vegetables, the vs human has a skeleto-muscular system that can be benefited by passive range of motion exercises. The vs human benefits from avoiding states that cause her to moan, scream, or cry incessantly. These states put stress on her heart, lungs and other bodily systems even if she cannot experience pain. These benefits are not simply benefits of a body or of a non-human animal, since animal identity is essential to being a human. If food and water bring stability to the body's system, they bring stability to the human being whose body is so benefited. Any bodily benefit is human benefit. ANH yields a genuine human benefit. The maintenance of a vs person's life with attention to maintaining muscles, avoiding contracture, releasing muscle tone by receiving passive exercise, and offering a calming aesthetically pleasing environment, all contribute to the well-being of the patient as a human. It is simply false to describe vs patients who lack external or internal awareness as able to engage only in vegetative operations. Vegetables don't have muscles or bones, so they don't require exercise or placement in standing frames to forestall osteoporosis. Vegetables don't have nervous systems or sensory organs, so they don't require an environment with pleasing colors, sounds, textures and aromas that contribute to the patient's stability and apparent satisfaction. Nor are vegetables benefited by human touch, hugs, grooming and bathing, as well as the soothing sound of a human voice speaking, praying or singing. The benefits of exercise, aesthetically pleasing environment and human interaction have been shown in studies that followed vs patients for a full year after onset of vs (Sazbon, 1990 and 2004). Food and water are necessary materials humans require to assimilate the goods of exercise, environment and human touch and voice. The good to be achieved in feeding and hydrating this human life, even in an unaware state are the goods humans share with other mammals, not simply the goods of vegetative life.

ANH makes the patient part of the community by continued participation in the life of the community. Feeding and being fed constitute community membership. In all cultures eating is an integral part of community-defining celebrations. Eating with others forms bonds and traditions. Proposals to marry, business deals, ideas for stories, paintings, music and scientific theories are hatched during shared meals. For Jews the Passover meal defines them as God's chosen people. For Christians, the central act of worship is a sacrificial meal of eating Christ's body and blood in the appearance of bread and wine. Feeding these people affirms them as our brothers and sisters. This feeding connects the patient to the community and thereby benefits the patient as ordinary care.

Some have argued that for many vs patients ANH can be judged an extraordinary care for the same reason a ventilator can be so judged (Ford, 2004). This is a misleading comparison. A ventilator breathes for a patient. ANH does not digest food for a patient. It simply delivers food and water in a way that the patient can receive it and digest it. Providing ANH to a patient is akin to providing oxygen for the patient to breathe. Providing oxygen, food and water are ordinary cares, not extraordinary medical treatments.

Many believe that the good achieved for the patient is disproportionate when the patient's benefit is living without a cognitive-affective life until death. In support of this belief they hold that the obligation to promote one's life is conditioned upon the good to be achieved. Even the good of being aware of pain and its absence is not a sufficiently human good to warrant the burden of providing ANH and the ensemble of accompanying cares.

While the flourishing of these bodily systems is not sufficient for a flourishing human life, they are constituent parts of such a life. Maintaining the excellence of the skeletal muscular and nervous and immune systems constitutes part of human well-being. Maintaining the sensory organs with pleasing sensory objects and human sensory stimulation contributes to the excellence of the senses. The practices of human touching, speaking and being present connects the patient to other humans and concretely connects the patient to the community. Even if unaware of being loved and cared for, the vs patient's quality of life is enhanced by such connection as evidenced by blood pressure, heartbeat, less brittle bones, less moaning and groaning, less blank or empty facial expressions.

If these goods are not sufficient to count as proportionate benefits, then newborns with Trisome 18 or static encephalopathy who may never experience the cognitive-affective life unique to humans would not realize goods sufficient for feeding them in assisted ways. The warrant for this obligation is that the practices of these ensemble of cares enhance goods that are constitutive parts of human flourishing. Any practice that enhances constitutive elements of human flourishing is sufficient to count as worth the *prima facie* burdens associated with delivering these practices. So the practices of providing exercises, calming and pleasing environments and human interactions are sufficient to count as worth the *prima facie* burdens the patient, family and community bear in delivering this care. A necessary condition for the patient receiving such goods is that the patient be fed and hydrated. The burden of feeding and hydrating the patient once the tube is inserted is not excessive. Thus, ANH of vs patients is not only ordinary, but also proportionate care.

The goods described above are significant for they constitute part of human flourishing. Humans in vs are owed these goods inasmuch as humans in vs possess a human nature, i.e. an orientation to a life of loving the good and knowing the truth within a human community. Evidence of this orientation's presence is seen in vs patients. The event related potential in 12% of vs and 76% of near-vs patients in response to semantically anomalous expressions is one kind of evidence. The smile that is likely a response to a greeting or a hug may be another. The ground for believing that such humans have this orientation is a metaphysical judgment: Necessarily, any human being possesses a human nature, i.e. a causal source for the unique capacities of this type of animal. The possession of human nature does not guarantee exercise of uniquely human capacities, since a human nature is an animal nature, a nature of a bodily being whose exercise of its unique capacities require as constitutive parts proper functioning bodily systems. Since these bodily systems can be damaged, many humans will not be able to exercise uniquely human capacities. That lacking does not entail that such humans lack a human nature. Evidence of the causal efficacy of this nature is seen in practices of brain repair or cognitive behavioral therapy that restore brain function necessary for the exercise of the uniquely human activities. Such repair is not possible with the brains of dogs and cats, for there is no human nature present. In addition, the belief that brain-damaged humans retain a human nature is part of what motivates work on cures and treatments which seek to restore the damaged bodily systems to health so an injured human can participate in uniquely human activities. These observations point to another good to be had for the vs human who is nourished and hydrated: possible future improved brain function which enables a vs patient to participate in activities approaching or realizing unique human capacities.

Another set of goods compensates the caregivers and community who initiate, sustain and deliver this care. MacIntyre points out how care for any sick or disabled member teaches the caregivers about the limits of our self knowledge. We learn to correct errors of practical judgment about others based on appearance, for example. We learn to value the courage and grace required from others who face this disablement which we ourselves may need at some later time (MacIntyre, 1999, p. 138). The lessons learned from caring for those human beings for whom the potentialities of rationality or affective response have been permanently frustrated merit MacIntyre's special attention.

> It will be urged that of such we can only say that at most they can be passive objects of benevolence designed to limit their suffering, beings whose existence can only be a cost, but in no way a benefit to others . . . Towards them I may adopt an attitude of benevolence, but my relationship to them must be one-sided. But this is a mistake. What they give us is the possibility of learning something essential, what it is for someone else to be wholly entrusted to our care, so that we are answerable for their well-being. Every one of us has, as an infant, been wholly entrusted to someone else's care, so that they were answerable for our well-being. Now we have the opportunity to learn just what it is that we owe to such individuals by learning for ourselves what it is to be so entrusted (MacIntyre, pp. 138–139).

This kind of entrusting is importantly different from the infant being entrusted to the parent. There may be no biological relationship between caregiver and patient. The infant, typically, has many potentialities for rational and affective activities unlike the vs patient. Consequently the entrusting relationship diminishes between parent and child, while it is likely to continue between the caregiver and vs patient until natural death. Thus, the lesson of this kind of responsibility requires more than the biologically rooted stirrings of care and affection found in new parents. Only those practiced in the virtues of acknowledged dependence towards profoundly disabled people can choose this kind of caring for its own sake. The rest of us can discover this insight once we have begun to do the actions required by such virtue. As Aristotle taught in the *Nicomachean Ethics* Book II, we become virtuous by doing the action a virtuous person does. As we develop the habit of responding in this way we see its inherent goodness and come to choose it for itself. With time and more practice we develop the character that marks one as having this virtue.

I have made a case for premise 2's truth. ANH for vs patients for whom death is not imminent and for whom the food and water can be assimilated without harming bodily systems necessary for survival is ordinary and proportionate care. ANH is not treatment aimed at curing or treating the damaged brain of the patient. While the insertion of a feeding and hydrating tube is a medical act, applying food and water through the tube is part of basic care owed any community member. If the community owes the patient the basic cares of clothing, shelter and comfort, it surely owes her the basic care of food and water. Since the brain-damage causing vs is not a terminal pathology and since there is no underlying lethal pathology which makes death imminent, the food and water is not simply forestalling the time of death. The food and water are necessary to sustain the vs patient's life as much as ANH supports the life of a patient with multiple sclerosis or Lou Gehrig's disease. In none of these patients does ANH cure or treat brain or nerve damage. Since ANH is morally obligatory for those afflicted with the neuro-muscular disease, it is morally obligatory for those enduring vs. The benefits of ANH for the vs patient are not simply the vegetative benefits of respiration, digestion and circulation of nutrients to body parts. The fact that the patient can realize goods constitutive of parts of human flourishing and that the cost of providing food and water through the tube is modest, confirm that the care is proportionate to the burden of providing it. If the cost of providing the other cares along with ANH for 2–10 plus years becomes excessive for the vs patient's family, then the community is obliged to ease that burden in the same way that it is so obliged for the families of patients with multiple sclerosis and Lou Gehrig's disease. The national cost of providing such care is not overwhelming (see Borgnovia, 2004).

Since these community supports and practices are not in existence now, an important consequence of this argument is the need for radical change in the way health insurance, doctors, nurses and health institutions and communities of all kinds work with vs patients. For families who have made a good will effort to seek community support in meeting the burdens of care beyond one year and who still face excessive economic or psychological burdens that disrupt obligations to themselves and other family members, this care can become disproportionate. Such families are relieved of their moral obligation to provide ANH to a vs patient. The basic ordinary care has genuinely become disproportionate to the benefit in such situations. The family intends cessation of a basic care that has become excessively burdensome to them. The resulting death by dehydration or starvation is foreseen but not intended. Such outcomes are tragic and should stir our communities to provide practices of supportive care that preclude such outcomes.

I turn to considering ways in which such practice can be implemented. These practices must be crafted at a community level above the family and below the state or nation, for the family lacks the expertise and financial resources and the state or nation lacks the shared commitment to the particular others required for exercising virtues of acknowledged dependence (MacIntyre, 1999, pp. 133–134). Neighborhood, parish or local professional organizations like the Lions Club, the Knights of Columbus, or scouting or professional groups are examples of the appropriate community level. I envision two models of support: (a) in-home supplemented with volunteers and bi-weekly visits by nurses and therapists, and (b) respite centers staffed by health professionals and volunteers.

The key to the viability of the in-home model are volunteer teams sponsored by local organizations cited above. Leaders within these organizations would establish teams of volunteers who would receive training in how to assist families caring for vs patients in their homes. These teams would regularly assist a family with day or evening time cares for the family's loved one. Physical therapy, occupational therapy and nursing professionals from local rehabilitation centers would educate and prepare family and team volunteers for the challenges and rewards of in-home care for vs patients. This would include learning about ways to meet the psychological, financial and social challenges of caring for such a person in one's home. Access to financing and health professionals would require state and national resources distributed through community-level organizations. The rehabilitation center would maintain contacts with organizations that supplied volunteer teams.

The key to the success of respite centers is private/public partnerships for building or remodeling facilities that provide environments for vs patients to experience regular quality human interaction in an environment that seeks a balance of proper sensory stimulation with calm and good quality custodial care. The work of trained professionals must be supplemented by volunteers in order to keep costs down. More importantly, there is the educative role of getting able-bodied people to work with vs patients to dispel the myth that these individuals are just vegetating or that they are in a coma or zombie-like state.

Some respite centers may be connected with University medical research centers where neurologists and doctors and other health care professionals can work with vs patients to learn more about the covert signs of awareness, seek ways of communicating with patients and ultimately learn how to restore the ability for voluntary activity in such patients.

Once there is a critical mass of vs patients and families who have had a positive experience caring for their loved one for more than a year using ANH, community support for such practices will grow. In time, caring for vs patients with ANH along with range of motion, pleasing environment and human interaction will become a common volunteer experience for high school and college students seeking to serve their community. Girl and boy scout organizations could earn badges in care for the brain damaged, for example. These kinds of activities support the exercise of the virtue of acknowledged dependence. The connections formed between the volunteer teams, the families and the health professionals will foster development of a more connected community, conscious of the challenges and rewards of embracing its own vulnerabilities.

## 2.5 Criticisms of the Pope's Argument

I turn to consider important criticisms of this moral obligation to feed vs patients. John Harvey, M.D. of the Georgetown University Center for Clinical Bioethics and Ron Hamel, Ph.D. of the Catholic Health Association of the United States make a case for the moral permissibility of withdrawing feeding tubes from vs patients in their paper, "On Withdrawing Medically Administered Nutrition and Hydration," presented at the Vatican conference on caring for vs patients (Hamel and Harvey, 2004). They offer two principal arguments. The first one is based on the Catholic

Church's tradition of not requiring all means of life support when the goal is mere survival.

> since this patient is in a persistent vegetative state with no hope of recovery, it would be morally permissible to withdraw the artificially administered nutrition and hydration. These interventions cannot restore the patient to health, improve her condition or afford her any other benefits except the prolongation of physical life. The latter can be the only goal or *telos* of continued artificial nutrition and hydration in this situation. Throughout our 500 year tradition relating to this matter, the mere preservation of vital functions has not been considered morally obligatory or sufficient in itself to oblige someone to use a particular means of preserving life (Hamel and Harvey, p. 5).

Their argument can expressed as a categorical syllogism modified to address ANH for the specific class of vs patients I am considering in this paper.

1. No medical treatments whose goal is only to sustain the physical life of a human lacking any future capacity for mental life are morally obligatory.
2. ANH for patients in vs for whom death is not imminent and for whom assimilation of food and water is possible without significant harm to bodily systems necessary for life is a medical treatment whose goal is only to sustain the physical life of a human lacking any future capacity for mental life.
3. So ANH for patients in vs for whom death is not imminent and for whom assimilation of food and water is possible without significant harm to bodily systems necessary for life is not morally obligatory.

They support the first premise by referencing a passage from Pope Pius XII in which he affirms that the temporal goods of life and health are subordinated to spiritual ends (Pius XII, 1957). They also cite Pope John Paul II from the encyclical, *Gospel of Life* where he affirms that in the obligation to care for oneself, one must take account of concrete circumstances, specifically, "whether the means of treatment available are objectively proportionate to the prospect for improvement" (John Paul II, 1995). Since the goal of sustaining nothing other than the physical life of a human is not proportionate to the unlikely recurrence of mental life, the goal of sustaining simply physical life is not obligatory. This argument challenges premise 2 of the Pope's argument.

A persuasive ground for denying this community obligation is that a member with no future capacity for mental life has no prospect of ever being able to flourish as a human being. Unlike an unborn child in its mother's womb who has a future capacity for mental life that grounds the mother's obligation to feed her child in utero, the vs patient has no realistic prospect of such capacity. Since the obligation of the community to provide common goods is rooted in the community's requirement to institute and sustain practices that provide conditions for its members' flourishing as human beings, and since these members have no present or future prospect for such flourishing, the community is under no obligation to deliver life-sustaining treatments to humans who will always lack this capacity for human flourishing.

A disturbing consequence of applying this principle to ordinary cares implies that the practices of dressing, sheltering, cleaning and grooming vs patients are not obligatory for they simply sustain the physical life of humans lacking any future capacity for mental states. While it may seem that premise 1 should be rejected since it entails a false consequence, Hamel and Harvey can rightly insist that premise 1 does not entail the false consequence, for what is described in the false consequences are not medical treatments, but ordinary cares. Since premise 1 ranges over medical treatments, not ordinary cares, premise 1 is not falsified by this consideration.

If this were Hamel's and Harvey's response, then ANH for vs patients does not fall under premise 1, since ANH is an ordinary care and not a medical treatment. ANH like dressing, grooming and sheltering profoundly ill patients are owed in virtue of the patients' membership in our community. These cares are required as exemplifications of the virtue of acknowledged dependence. These are good reasons for denying premise 2's truth.

Furthermore, I have argued that providing ANH is a proportionate good for the vs patient since it is necessary for the patient to receive the goods of exercise, aesthetically pleasing environment and human interaction which constitute parts of a humanly flourishing life. Hamel and Harvey confuse matters when they abstract from the lived unity of human life to speak of mere physical life distinguished from mental life. Since humans are essentially animal organisms, flourishing in the exercise of their neuromuscular systems or nervous and immune systems are constitutive parts of genuine human flourishing. Hence premise 2 is false, for ANH is a care not a treatment, and it is a care constitutive of human flourishing.

Premise 1 suffers from an impoverished human anthropology. Humans are not composed of mental and physical lives in which only humans with a mental life ground moral obligations for cares and treatments. Humans are animal organisms whose flourishing as humans includes non-mental activities as well as mental ones. While it may be true that no medical treatment or care that contributes nothing to human flourishing is morally obligatory, it is without a doubt true that ANH to vs patients can and does contribute to human flourishing.

Harvey's and Hamel's second argument challenging the obligation to feed vs patients rests on their interpretation of the meaning of "reasonable hope of benefit" for the patient. They write,

> for traditional moralists, a reasonable hope of benefit was not limited to sustaining life ... Rather the meaning of benefit was relative to the condition of the patient and included at least restoration to health or improvement in one's condition, relief of pain and maximization of comfort (p. 5).

Their argument can be expressed as a *modus tollens*:

1. If a treatment is morally obligatory then the treatment must offer a reasonable hope of benefit to the patient.
2. Some practices of feeding vs patients who can receive life-sustaining nourishment do not offer a reasonable hope of benefit.
3. So some practices of feeding vs patients who can receive life-sustaining nourishment are not morally obligatory.

Harvey and Hamel define "reasonable hope of benefit" as restoring patient to health, improving their condition, relieving their pain or maximizing their comfort. With this definition of "reasonable hope of benefit" premise 2 is true. However using that definition renders premise 1 false, for with that definition there would be no moral obligation to feed individuals with multiple sclerosis or Lou Gehrig's disease who are unable to feed themselves. In the later stages of multiple sclerosis or Lou Gehrig's disease patients are often unable to swallow or chew. ANH for such patients fails to meet the conditions of reasonable benefit for it neither restores these individuals to health, nor improves their condition, nor maximizes their comfort. The food does however sustain their life. While such feeding prevents the pain of starvation, no physician considers food pain prevention medicine. Thus, this narrow interpretation of reasonable hope of benefit cannot be correct. Neither criticism of the Pope's argument succeeds in undermining the case for the moral obligation of the community to provide ANH to this class of vs patients.

In this paper I have offered a philosophical defense of Pope John Paul II's teaching that ANH for vs patients for whom death is not imminent and for whom assimilation of food and water is possible without significant harm to bodily systems necessary for life is morally obligatory for the community to provide and the patient to accept. The obligation is grounded on two premises: Any ordinary and proportionate care for patients is morally obligatory and ANH for such patients is ordinary and proportionate care. I have supported the first premise based on the obligation of the community to seek the common good of its members. On one account, if these cares are not morally obligatory, then the community fails to aim at the good of each of its members and so loses its moral legitimacy to exercise authority over all its members. On the other account, if these cares are not obligatory, the conditions required for humans realizing their common good are not met. I supported the second premise based on the benefits the vs patient can realize from the ensemble of cares made possible by the continued life which ANH makes possible. These benefits are constitutive parts of human flourishing. In addition, providing this care forges a solidarity among the able-minded and brain-injured that contributes to the moral development of the sustainers and practitioners of this care. The community's role in institutionalizing this practice is necessary for families to be able to provide these benefits to their loved ones in the vs state. In refusing to cut off these members from participating in the common life of the community, patients and caregivers are tethered more closely, realizing each in their appropriate ways a humanly flourishing community.

## Notes

[1] R. Doerflinger writes that the Pope's teaching coheres with the 1992 statement (Doerflinger, 2004), as does Kevin McMahon (McMahon, 2004).
They note that the US Bishops held that the presumption was to be in favor of giving food and water and that the Bishops had rejected the view that since the vs patient can no longer engage in prayer and spiritual the burden of artificial feeding was not worth the benefit of continued merely physical existence. Neither of them explicitly addresses the Bishops' explicit endorsement of withdrawal of ANH for excessive burden it may place upon the patient or family or community.

[2] Absent community support individual families may rightly judge that the burden is excessive, i.e. more than they can afford economically or psychologically. This is why the Pope's message calls for social change in practices of caring for such patients. Community support will require changes in the health care system and medical and pastoral practices.

[3] See Shannon and Walter (2004); Hamel and Michael (2004); Tuohey (2004); Coleman (2004).

[4] There is agreement on the general causes of the condition. It results from severe injury to the brain due to trauma like a car accident, or a degenerative disease like Alzheimer's. Some of the degenerative diseases are associated with chronic brain injury which are not preceded by comas. The acute brain injuries are usually preceded by a coma which after 2–4 weeks result in recovery of awareness, a vegetative state or death. The vegetative state itself may end in recovery of full consciousness after days or months, or it may persist until death (Sazbon, 1990, 2004). Three different population studies have shown that after one year 52 % of vs have recovered, 33 % have died, and 15 % continue in vs. After 3 years another 2 % of the total recover. Recoveries after 3 years occur rarely (Multi-Society Task Force on PVS (MSTF), pp. 1572, 1575–1577 & Sazbon, 1990, 2004).

[5] The professional societies include the American Academy of Neurology, Child Neurology Society, American Neurological Association, American Association of Neurological Surgeons, and American Academy of Pediatrics.

[6] Shewmon also makes this point in (Shewmon, 2004b, p. 345). The role of thalamus in pain is described in Willis (1989), the modulatory role of the cortical areas is described in Bonica (1990).

[7] Another researcher recorded spectral EEG readings in vs patients receiving anesthesia or dental surgery. The EEG described spectral EEG monitoring in vs patients undergoing anesthesia for dental surgery (See Pandit et al., 2002). The EEG changes were similar to those in neurologically normal patients undergoing the same procedures. These authors concluded: "The findings raise the possibility that patients in permanent vegetative states might sense noxious stimuli at a cortical level."

[8] The task force recognizes that the "locked in" state is an exception to this inference. In such a condition the patient's brain is fully functional, but the patient is unable to move any facial or bodily muscles to communicate awareness. The task force holds that this possibility is sufficiently rare that it does not interfere with a clinical diagnosis carefully established by experts (MSTF, p. 1501). Later they indicate that brain imaging and positron-emission tomography can be used to distinguish the two states, (MSTF, p. 1502).

[9] A.A. Howsepian presented a strong critique of the MSTF's judgment that all vs patients are unconscious. In light of the discoveries reported in the paper, his logical critique was most prescient. See Howsepian, 1996.

[10] Giacino underlines that this prognosis is not absolute by citing three factors. First, the MSTF had evidence of only 53 survivors beyond one year and half of the records for that number were anecdotal. Second, there are professionally peer-reviewed documented recoveries of awareness. Finally the evidence is on patients treated twenty years ago when medical knowledge of the neurological conditions of such patients was not as advanced as it is today (Giacino, 2004, p. 295).

[11] This is the preference of B. Jennet, co-originator of term 'vegetative state' as reported in personal communication received in 1995 by Nathan Zasler as reported in 'Terminology Evolution: Caveats, Conundrums and Controversies,' *Neurorehabilitation* 19 (2004): 289. In addition, it was the practice used by the World Federation of Catholic Medical Associations and the Pontifical Academy for Life in their joint statement released on March 20, 2004, titled "Considerations on the Scientific and Ethical Problems Related to Vegetative State." See http://www.vegetativestate.org/documento_FIAMC.htm

[12] The prognosis of permanent vs is given to traumatic brain injured after 12 months, to the non-traumatic brain injured after 3 months, to the congenitally malformed after 3–6 months, to those with degenerative diseases after 1–3 months (Giacino, 2004, p. 295).

[13] I use good of a human and good for a human as Aristotle does in Nicomachean Ethics 1.7. The good of a human is the excellent exercise of a human's characteristic activities while the good for a human is any benefit that accrues to the human.

[14] MacIntyre is strong on this point, "My regard for another is always open to be destroyed by what the other does, by serious lies, by cruelty, by treachery . . . but if it is diminished or abolished

by what happens to the other, by her or his afflictions, then it is not the kind of regard necessary for those communal relationships-including relationships to those outside the community-through which our common good can be achieved" (MacIntyre, 1999, p. 128).

# References

American Congress of Rehabilitation Medicine. (1995). Recommendations for use of Uniform Nomenclature Pertinent to Persons with Severe Alterations in Consciousness, *Archive of Physical Medicine Rehabilitation 76*, 205–209.

American Medical Association. (1986). Council on ethical and judicial affairs. *Opinion 2*, 15.

Andrews, K. (1992). Letting vegetative patients die, (Letter). *British Medical Journal 305*, 1506.

British Medical Research Council. (1973). Report of the board of science and education. *Medical Tribune 19*, 5.

Beuthien-Baumann, B., Handrick, W., Schmidt, T., et al. (2003). Persistent vegetative state: Evaluation of brain metabolism and brain perfusion with PET and SPECT. *Nuclear Medical Community 24*, 643–649.

Bonica, J. J. (1990). Anatomic and physiologic basis of nociception and pain. In J.J. Bonica (Ed.), *The Management of Pain* (Vol. 1), Lea & Febiger: Philadelphia, pp. 28–94.

Borgnovia, E. (2004). Economic aspects in prolonged life sustainable treatments. *Neurorehabilitation, 19*, 367–371.

Coleman, G. (2004). Take and eat: Morality and medically assisted feeding. *America 190*(12), 16–20.

Doerflinger, R. (2004). Pope John Paul II affirms obligation to feed patients in the "Vegetative" state. *Ethics and Medics 29*(6), 2–4.

Ford, N. (2004). Determining what is best for patients. *Origins, 33*(43), 751–752.

Giacino, J. T., Zasler, N. D., Katz, et al. (1997). Development of practice guidelines for assessment and Management of the Vegetative and Minimally Conscious States. *Journal of Head Trauma Rehabilitation 12*(4), 79–89.

Giacino, J. T., Ashwal, S., Childs, N., et al. (2002). The minimally conscious state: Definition and diagnostic criteria. *Neurology 58*, 349–353.

Giacino, J. T. (2004). The vegetative and minimally conscious states: Consensus-based criteria for establishing diagnosis and prognosis. *NeuroRehabilitation 19*, 293–298.

Gigli, G. L., & Valente, M. (2004). The withdrawal of nutrition and hydration in the vegetative state patient: Societal dimension and issues at stake for the medical profession. *NeuroRehabilitation 19*, 315–328.

Hamel, R., & Harvey, J. (2004). On withdrawing medically administered nutrition and hydration. *Origins 33*(43), 748–751.

Hamel, R., & Michael, P. (2004). Must we preserve life? The narrowing of traditional catholic teaching. *America 190*(14), 6–13.

Howsepian, A. A. (1996). The 1994 Multi-society task force consensus statement on the persistent vegetative state: A critical analysis. *Issues in Law & Medicine 12*(1), 3–29.

John Paul II. (1995). *Evangelium Vitae*. Vatican City: Libreria Editrice Vaticana.

John Paul II. (2004). Care for patients in the permanent vegetative state. *Origins 33*(43), 738–740.

Kassubek, S., Juengling, F. D., Els, T., et al. (2003). Activation of a residual cortical network during painful stimulation in long term post-anoxic vegetative state: A 15O-H2O PET study. *Journal of Neurological Science 212*, 85–91.

Kotchoubey, B., Lang, S. V., Bostanov et al. (2002). Is there a mind? Electrophysiology of unconscious patients. *News Physiological Scientist 17*, 38–42.

Laureys, S., Faymonville, M. E., Luxen, A., et al. (2000). Restoration of thalamocortical connectivity after recovery from persistent vegetative state. *Lancet 355*, 1790–1791. [Erratum in: Lancet 2000; 1355(9218):1916].

Laureys, S., Faymonville, M. E., Peigneux, P., et al. (2002). Cortical processing of noxious somatosensory stimuli in the persistent vegetative state. *Neuroimage 17*, 732–741.

MacIntyre, A. (1999). *Dependent Rational Animals*. New York: Open Court.

McMahon. K. T. (2004). Catholic moral teaching, medically assisted nutrition and hydration and
the vegetative state. *Neurorehabilitation 19*, 373–379.

Multi-Society Task Force on PVS. (1994). Medical aspects of the persistent vegetative state (Part I).
*New England Journal of Medicine 330*(21), 1499–1508.

Multi-Society Task Force on PVS. (1994). Medical aspects of the persistent vegetative state
(Part II). *New England Journal of Medicine 330*(22), 1572–1579.

Pandit, J. J., Schmelzle-Lubiecki, B., Goodwin, M., et al. (2002). Bispectral index-guided management of anesthesia in permanent vegetative state. *Anesthesia 57*, 1190–1194.

Pius XII. (1957). On the prolongation of life. *The Pope Speaks 4*(4), 393–398.

Pro-Life Activities Committee of the United States Conference of Catholic Bishops. (1992). Nutrition and hydration: Moral and pastoral reflections. *Origins 21*(44), 705–712.

Sazbon, L., Wasserman, Z. (1990). Outcome in 134 patients in posttraumatic unawareness
(Part 1): Parameters determining recovery of consciousness. *Journal of Neurosurgery 72*,
75–80.

Sazbon, L. (2004). Rehabilitation of patients in vegetative state. *L'Arco Di Giano 39*, 108–116.

Schiff, N. D., Ribary, U., Moreno, D. R., et al. (2002). Residual cerebral activity and behavioral
fragments can remain in the persistently vegetative brain. *Brain 125*, 1210–1234.

Schoenle, P. W., Witzkey, W. (2004). How vegetative is the vegetative state? Preserved semantic
processing in vs patients – Evidence from N 400 event-related potentials. *NeuroRehabilitation
19*, 329–334.

Shannon, T., & Walter, J. J. (2004). Implications of the papal allocution on feeding tubes. *Hastings
Center Report 34*(4),18–20.

Shewmon, D. A. (1997). Recovery from "Brain death": A neurologist's apologia. *The Linacre
Quarterly 64*(1), 59–60.

Shewmon, D. A., Holmes, G. L., & Byrne, P. A. (1999). Consciousness in congenitally decorticate
children: Developmental vegetative state as self-fulfilling prophecy. *Developmental Medical
Child Neurology 41*, 364–374.

Shewmon, D. A. (2004a). 'The ABC of PVS: Problems of definition. In C. Machado, &
Shewmon, D. A. (Eds.), *Brain Death and Disorders of Consciousness*, New York: Kluwer
Academic/Plenum Publishers, 215–228.

Shewmon, D. A. (2004b). A critical analysis of conceptual domains of the vegetative state: sorting
fact from fancy. *NeuroRehabilitation 19*, 343–347.

Tuohey, J. (2004). The Pope on VS: Does JP II's statement make the grade? *Commonweal 131*(12),
10–13.

Willis, W. D. (1989). The origin and destination of pathways involved in pain transmission. In P.D.
Wall, & Melzack, R. (Eds.), *Textbook of Pain* Churchill, Livingstone, Edinburgh, 112–127.

# Chapter 3
# Caring for Persons in the "Persistent Vegetative State" and Pope John Paul II's March 20 2004 Address "On Life-Sustaining Treatments and the Vegetative State"*

William E. May

After giving a brief "historical background" to my involvement in this issue, I will summarize responses from various sources to the question of providing food/hydration by tubal means to persons alleged to be in this condition prior to Pope John Paul II's March 20, 2004 address; in particular, I will examine and criticize a major argument advanced by Catholic theologians to support the view that such provision of food/hydration constitutes extraordinary or disproportionate treatment and is hence not morally obligatory and then present the reasoning given by others, Catholic and non-Catholic, to support the claim that such provision is morally required and must be regarded as ordinary or proportionate means of preserving life. I will then consider Pope John Paul II's address, some negative reactions it has received, and a defense of it. I will then offer a conclusion.

## 3.1 Historical Background

In my book *Human Existence, Medicine and Ethics: Reflections on Human Life*, published in 1977, I claimed, in commenting on the Karen Quinlan case, that there was no obligation "to use the means currently employed to prolong her death" and that it would be morally permissible for her parents and others to "remove the tubes necessary for her feeding, prevent dehydration by appropriate medical means, and attend to her in her dying moments" (May 1977, 150–51). But in 1985 a study group of Pontifical Academy of Sciences released a report on the artificial prolongation of life. In it were included "Medical Guidelines" which declared: "If the patient is in

William E. May
Pontifical John Paul II Institute for Studies on Marriage and Family,
Catholic University of America
Email: wmay@johnpaulii.edu

* This essay originally appeared in *Medicina e Morale* 55 (Maggio/Giungno 2005) 535–555, and is reprinted here, with minor additions that bring it up to date, with permission.

a permanent, irreversible coma, as far as can be foreseen, treatment is not required, but all care should be lavished on him, *including feeding*" (emphasis added) (1985). I thus began to reconsider the position I had taken in 1977.

In 1986 Rev. William Gallagher, then president of the Pope John XXIII Center (now the National Catholic Bioethics Center), asked me to convene a group of moral theologians, moral philosophers, lawyers, medical doctors, and nurses to study the issues involved in providing food and hydration to the permanently unconscious and other vulnerable persons. I was then teaching at The Catholic University of America and I succeeded in having the following persons meet several times over several weeks to study the issue and to prepare a final paper: Benedict Ashley, O.P., Robert Barry, O.P., Msgr. Orville Griese, Germain Grisez, Brian Johnstone, C.Ss.R., Thomas Marzen, J.D., Bishop James McHugh, S.T.D., Gilbert Meilaender, Ph. D., Mark Siegler, M.D., and Msgr. William Smith. Some nurses who had spent hours caring for individuals in the so-called persistent vegetative state also attended the meetings.

We learned that individuals in this condition are *not* suffering from a fatal pathology and that they are in relatively stable condition and capable of living for some time so long as they receive food and hydration. We learned that at the beginning they are capable of swallowing, but that feeding them orally takes a great deal of time and that using tubes to feed them lightens the burdens of their care-givers. We also learned that the cost of feeding them is very reasonable, and that they do not have to be kept in expensive institutions but can be cared for at home if there is someone there to provide care who can be helped by visiting nurses, etc.

As a result of the new knowledge two of us who had previously thought that tubally providing such persons with food and nutrition constituted extraordinary care and hence could be morally omitted, changed our minds—Germain Grisez (1986) and I. Benedict Ashley, another member of the group who had previously judged it not morally required to provide food and hydration by tubal means, did not change his mind and accordingly did not sign the final paper prepared and approved by the other participants in the meetings and subsequently published in the journal *Issues in Law & Medicine* under the title "Feeding and Hydrating the Permanently Unconscious and Other Vulnerable Persons" (May, et al. , 1987). I will later summarize the reasoning advanced in this paper.

## 3.2 Responses by Various Sources to the Issue Prior to Pope John Paul II's Address

Here I consider (1) recommendations of some professional bodies and bioethics centers; (2) major responses by bishops of the United States; (3) a summary and critique of the major argument advanced by Catholic theologians who claim that such feeding is not obligatory; and (4) a summary of the argument given by those who maintain that such feeding is morally obligatory unless it can be clearly shown to be futile.

## 3.2.1  Recommendations by Professional Bodies, etc.; PVS Patients, Consciousness, and Pain

During the 1980s and 1990s court cases involving termination of tubal feedings of PVS patients proliferated, eliciting responses from various professional organizations. In 1981 the Judicial Council of the American Medical Association declared it ethical to withdraw all means of life support, including such feeding, "where a terminally ill patient's coma is beyond doubt irreversible" (1981, p. 9). The President's Commission for the Study of Ethical Problems in Medicine and Biomedical and Behavioral Research addressed this issue in its 1983 monograph on foregoing life-sustaining treatments; it concluded that the decision to provide or forego tube feeding of PVS patients was best made by the patient's surrogates and not by the courts. It likewise concluded that foregoing all treatment, including tubally administered food and hydration, was an ethically legitimate option (1983, pp. 171–96). Other organizations issuing guidelines favoring the withholding of tubal feeding from PVS patients the Hastings Center (1987), the American Academy of Neurology (1989, pp.125–26), and the American Medical Association (1990, 426–30). It should be noted that a significant number of the individuals who drafted statements of this kind think that "personhood" is lost if an individual is no longer capable of exercising cognitive abilities.

More importantly, as D. Alan Shewmon, M.D., himself a leading neurologist and an expert particularly on the neurology of the brain, has noted, the unquestioned acceptance among medical authorities that patients with widespread cortical damage are ipso facto unconscious and incapable of experiencing pain and suffering is not based on verifiable evidence but is accepted "because official neurology says so." Shewmon concluded that upon critical examination the "evidence" alleged to support the claim that patients with widespread cortical damage are by definition unconscious and incapable of feeling pain is "of an exclusively negative nature: patients with diffuse cortical destruction do not manifest clinical signs of awareness of self or environment. But there was no positive evidence that such patients are not inwardly conscious." Continuing, he observed that "no one seemed concerned that perhaps what is eliminated by cortical destruction might be the capacity for external manifestation of consciousness rather than consciousness itself; in other words, that what is called "PVS" might in reality be merely a "super-locked-in" state [a condition in which the person is indeed conscious but is utterly incapable of manifesting this externally]" (1997, 64.1, 59–60). Shewmon then went on to say the following, in a passage of single importance with respect to the question of pain experienced by PVS patients:

> The more I reconsidered the matter, the more I began to realize that the supposed lack of evidence for consciousness was not even complete. For example, all treatises on the neurophysiology of pain traced the anatomical pathway from the cutaneous nociceptors centrally, invariably ending not at the cortex but at the thalamus. Patients with strokes involving somatosensory cortex lose tactile discrimination and joint position sense, but not the capacity to perceive and to localize pain. Thalamic injury, however, can cause a distressing form of central pain. In the pain literature it is clear that the cortex's role in pain perception is merely

modulatory and that the experience is mediated subcortically, but in the PVS literature these well known phenomena are systematically ignored. PVS patients often grimace to noxious stimuli and manifest primitive withdrawal responses. Advocates of the cortical theory write off such behaviors as mere brain-stem or spinal reflexes, but that dismissive attitude is based more on an apriori assumption than a scientific conclusion (1997, p. 60).

## 3.2.2 Responses by U.S. Bishops

Prior to John Paul II's address in March 2004 the universal Magisterium of the Church had not specifically addressed this question, while the bishops of the U.S. had given contradictory answers. Directive no. 58 of the *Ethical and Religious Directives for Catholic Health Care Services* (Nov. 1994) holds that one ought to presume that nourishment so provided be given such persons "as long as this is of sufficient benefit to outweigh the burdens involved to the patient." Despite this, however, some individual bishops and the Texas Conference of Catholic Bishops concluded that providing "food" through tubal means is futile and useless. The Texas Bishops, who did not provide extensive argument, believed that someone in PVS was "stricken with a lethal pathology which, without artificial nutrition and hydration will lead to death." They held that withholding or withdrawing artificially provided food from such persons "is simply acknowledging the fact that the person has come to the end of his or her pilgrimage and should not be impeded [by artificially provided food] from taking the final step." In short, the Texas bishops judged such provision of food futile or useless and hence not obligatory (1990). Some individual bishops, e.g., Louis Gelineau, et al. issued statements of a similar nature (1988, pp. 546–47).

On the other hand, on March 24, 1992 the Administrative Committee of the National Conference of Catholic Bishops authorized the publication of a substantive document prepared by the Committee for Pro-Life Activities of the NCCB. This document surveyed somewhat extensively relevant medical literature dealing with the issue and different positions taken by moral theologians. In their review of theological opinions, the authors of this document explicitly state that they do not find persuasive the rationale of some theologians that since persons in the PVS condition can no longer pursue the spiritual goal of life feeding them artificially is futile and unduly burdensome. In the conclusion of their paper, the authors have this to say: "We hold for a presumption in favor of providing medically assisted nutrition and hydration to patients who need it, which presumption would yield in cases where such procedures have no medically reasonable hope of sustaining life or pose excessive risks or burdens" (1992, p. 7).

It is noteworthy that John Paul himself singled out this paper for praise in a talk to a group of U.S. bishops on their *ad limina* visit to the Vatican in 1998:

The statement of the U.S. bishops' pro-life committee, 'Nutrition and Hydration: Moral and Pastoral Considerations,' rightly emphasizes that the omission of nutrition and hydration intended to cause a patient's death must be rejected and that, while giving careful consideration to all the factors involved, the presumption should be in favor of providing medically assisted nutrition and hydration to all persons who need them (1998, no. 4).

The Pennsylvania bishops had issued shortly before a somewhat similar document, replete with references to pertinent medical literature, on January 14, 1992. In its conclusion the bishops declared: "As a general conclusion, in almost every instance there is an obligation to continue supplying nutrition and hydration to the unconscious patient. There are situations in which this is not the case [e.g., when the patient can no longer assimilate the food and its provision is hence useless], but these are exceptions and should not be made into the rule." In their judgment artificially providing food to PVS patients is "clearly beneficial in terms of preservation of life," nor does it, in almost every case, add a "serious burden." Consequently, it is morally obligatory (1992, pp. 542–53).

Several individual bishops, e.g., James McHugh, Bishop fo Camden, N.J pp. 314–15), and other conferences of bishops, e.g., the New Jersey State Catholic Conference (1987, 542–553), issued statements reaching similar conclusions as the Pro-Life Committee and the Pennsylvania bishops.

### 3.2.3 The Theological Position Claiming that Tubal Feeding of PVS Patients is Not Morally Required

The leading proponent of this position prior to John Paul II's March 20, 2004 address, was Kevin O'Rourke, O.P., who presented his position in several places from 1986 through 2001 (1986, 321–33; 1988, 28–35; 1989, 181–96; 1989, 351–52; 1997, 421–26; 1999, 2001, 68.3, 201–17). O'Rourke based his claim that it is not morally obligatory to provide food/hydration to the permanently unconscious in the teaching of Pope Pius XII. In an important address to a congress of anesthesiologists, Pius had this to say:

> normally one is held to use only ordinary means [to prolong life] – according to the circumstances of persons, places, times, and culture – that is to say, means that do not involve any grave burdens for oneself or another. A more strict obligation would be too burdensome for most men and would render the attainment of the higher, more important good too difficult. Life, health, all temporal activities are in fact subordinated to spiritual ends. On the other hand, one is not forbidden to take more than the strictly necessary steps to preserve life and health so long as he does not fail in some more important duty (1957).

O'Rourke claimed that the Pope's emphasis on the spiritual goal of life specifies more clearly the terms "ordinary" and "extraordinary." A more adequate and complete explanation of "ordinary" means to prolong life would be: those means which are obligatory because they enable a person to strive for the spiritual purpose of life. "Extraordinary" means would seem to be those means which are optional because they are ineffective or a grave burden in helping a person strive for the spiritual purpose of life (1988, p. 32).

O'Rourke correctly interprets the teaching of Pius XII when he says that a means is extraordinary if it *imposes* a grave burden on a person and prevents him from pursuing the spiritual goal of life. But he errs greatly when he claims that a means is extraordinary when it is "*ineffective . . .* in helping a person strive for the spiritual purpose of life" and that a means is ordinary precisely because it enables a person to strive for the spiritual purpose of life. Many people, including some seriously handicapped children and mentally impaired adults, are incapable of pursuing the spiritual goal of life. They cannot do so because in order to do so a person must be able to make judgments and free choices. But these unfortunate human beings are still persons; their lives are still good, and it is good for them to be alive. If they should fall sick or be otherwise in danger of death, they surely have a right to "ordinary" care, and others have a serious moral responsibility to protect and preserve their lives unless the efforts to do so are futile or excessively burdensome. Thus, for example, if an elderly person suffering from a malady that renders him incompetent and incapable of engaging in human acts should suffer a cut artery and be in danger of dying because of loss of blood, it would surely be morally obligatory to stop the bleeding by appropriate means. Such means are surely "ordinary" or "proportionate." Yet, on O'Rourke's analysis, they would be "extraordinary" inasmuch as they would in no way be *effective* in helping this person to pursue the spiritual purpose of life.

It is instructive to note that Charles E. Curran, in a recent book highly critical of the moral theology of John Paul II, appeals to the same text of Pius XII in order to claim, falsely, that unlike Pius XII, John Paul II has absolutized bodily integrity (2005, p. 114).

Applying his understanding of Pius XII's teaching to persons said to be in the PVS condition, O'Rourke maintained that since these individuals are not capable of pursuing the spiritual goal of life and since feeding them tubally is ineffective in helping them do so, then such feeding is not required. He maintained, in addition, as did the Texas Bishops, that such individuals are suffering from a "fatal pathology." He likewise claimed that all one does by "feeding" such persons tubally is to preserve "mere physiological functioning." His associate Benedict Ashley, O.P., shared this position (2002, pp.421–22). A view similar to that taken in 2002 by Ashley and O'Rourke is that of James Walter and Thomas Shannon (1988). I believe that this view at least tends toward dualism if it is not in essence dualistic, since he claims that all one preserves by providing food to such persons is mere *physiological* functioning. I think that what one preserves is the *life* of those persons, who are not, as he contends, suffering from a fatal pathology such as cancer or congestive heart disease. I am happy to note that, in the recently published (December 2006) 5th edition of their *Health Care Ethics*, written with a new co-author, Jean Dublois, Ashley and O'Rourke accept the teaching of John Paul II's Address of March 20, 2004.

### 3.2.4 The Position Holding that Artificially Providing Food to PVS Persons is Obligatory

This view, like that of the Pennsylvania Bishops and the Pro-Life Committee of the National Conference of Catholic Bishops, holds that artificially providing food to permanently unconscious persons (those in the PVS state) is to be regarded ordinarily as morally obligatory insofar as it is neither useless nor unduly burdensome.

This was the position I and my collaborators—Robert Barry, O.P., Orville Griese, Germain Grisez, Brian Johnstone, C.Ss.R., Thomas J. Marzen, Bishop James T.McHugh, Gilbert Meilaender, Mark Siegler, M.D., and William B. Smith—developed in the article referred to earlier in this paper.

We began by articulating major presuppositions and principles, among them (1) that human bodily life is a great good, that it is personal, not subpersonal, that it is inherently good not merely instrumentally so, that no matter how heavily burdened such life remains a good; (2) that human life, however heavily burdened, remains a good of the person and that remaining alive is never rightly regarded as a burden and deliberately killing innocent human life is never rightly regarded as rendering a benefit.

We held that withholding/withdrawing various forms of preserving life, including the provision of food and water by tubal means, is morally permissible if the means employed is either useless or excessively burdensome. We held that it is useless or relatively so if the benefits provided are nil or insignificant in comparison to the burdens imposed, and that it is excessively burdensome if benefits offered are not worth pursuing for one or more objective reasons: too painful, too damaging to the person's bodily life and functioning, too restrictive of the patient's liberty and preferred activities, too suppressive of the person's mental life, too expensive, etc.

We acknowledged explicitly that "*if it is really useless or excessively burdensome* to provide someone with nutrition and hydration, then these means may rightly be withheld or withdrawn, *provided* that this omission does not carry out a proposal to end the person's life but rather is chosen to avoid the useless effort or the excessive burden of continuing to provide the food and fluids" (1987, 209). However, after examining the issue carefully, we judged that tubally providing food and hydration to the permanently unconscious and other vulnerable persons was neither useless nor excessively burdensome and that consequently it ought to be given. We thus concluded that:

> in the ordinary circumstances of life in our society today, it is not morally right, nor ought it to be legally permissible, to withhold or withdraw nutrition and hydration provided by artificial means to the permanently unconscious or other categories of seriously debilitated but nonterminal persons. Rather, food and fluids are universally needed for the preservation of life, and can generally be provided without the burdens and expense of more aggressive means of supporting life. Therefore, both morality and law should recognize a strong presumption in favor of their use (1987, 211).

We also argued that by caring for such persons in this way another good, that of human solidarity, was served. This point has been further developed by one of the co-authors, Germain Grisez who in a later work has said: "life-sustaining care for [persons] severely handicapped does have a human and Christian significance in addition to the one it would derive precisely from the inherent goodness of their lives. This additional significance is . . . profoundly real, just as is the significance of [a husband's faithfulness to a permanently unconscious] wife, which continues to benefit not only the person being cared for but the one giving care" (1997, p. 233).

Some, for example O'Rourke, argued that the expense entailed in feeding PVS patients must realistically be regarded as terribly burdensome in our society. He posed a question in his "Open Letter to Bishop McHugh" (who had held that the expense of feeding such persons is not excessive) that merits response. He said that McHugh was "disingenuous" in saying that "assisted nutrition and hydration . . . are not overly expensive" in view of the fact that "care in a hospital or long-term care facility costs anywhere from $600 to $1300 a day . . ." (1989, 351–52).

No one, we can presume, would want his family bankrupted in order to provide him with tubally assisted feeding. But does this mean that this claim is correct? At present, the cost for taking care of PVS patients is usually covered in great part by insurance or other programs. But one can legitimately avoid excessive expense (if this does become an issue) *without abandoning care for the person and without bringing his death about by starvation.* Persons put into the situation of caring for a loved one in the PVS state or other conditions are not obliged to have them cared for in highly expensive hospitals or nursing homes (if insurance and governmental help are inadequate). They can remove them from these costly institutions, take them home and do the best they can with the help of such services as hospice care, volunteers from the parish or neighborhood, etc. The high standards of care possible in expensive institutions might not be possible, but one can still maintain solidarity with the person doing what one can, including providing food and nourishment by tubal means (not too difficult to do once begun, even at home). One does not have to endure undue financial burdens.

## 3.3 Pope John Paul II's Address of March 20, 2004

### 3.3.1 Context and Key Themes

The Holy Father's Address came at the conclusion of an international congress entitled "Life-Sustaining Treatments and Vegetative State: Scientific Advances and Ethical Dilemmas," co-sponsored by the Pontifical Academy for Life and the International Federation of Catholic Medical Associations. The work of this group is well summarized by Richard Doerflinger, a participant at the Congress (2004, 3). The Pope's address was thus based on latest medical and scientific findings relevant to the "vegetative" state. These showed that this "state" is frequently misdiagnosed, that prognoses are far from reliable, and that the assumption that this state involves complete absence of unresponsiveness can be seriously questioned. In addition, scientific papers presented at the congress showed clearly that individuals said to be in the "persistent vegetative state" are not suffering from any fatal pathology or underlying disease that can cause their death, although they will, of course, die of dehydration if they do not receive food and water. It should be noted, too, that even prior to this congress medical doctors acknowledged that persons in this condition can usually be fed orally at the beginning, but that feeding them by tubal means is far more convenient to their care-givers and is more efficient and that, if not fed orally, the ability of persons in this state to take food orally gradually atrophies (1987, 904–11).

Among the principal themes developed by John Paul II in his Address are the following:

1. *"A man, even if seriously ill or disabled in the exercise of his highest functions, is and always will be a man,* and he will never become a 'vegetable' or an 'animal.' " (no. 3; emphasis in the original).
2. *The right of the sick person, even one in the vegetative state, to basic health care.* Such care includes "nutrition, hydration, cleanliness, warmth, etc." and "appropriate rehabilitative care" and monitoring "for clinical signs of eventual recovery" (no. 4).
3. *The moral obligation,* in principle, *to provide food and water to persons in the "vegetative" state by tubal means:* "I should like to underline how the administration of food and water, even when provided by artificial means, always represents a *natural means* of preserving life, not a medical act. Its use, furthermore, should be considered, in principle, *ordinary and proportionate,* and as such morally obligatory, insofar as and until it is seen to have attained its proper finality, which in the present case consists in providing nourishment to the patient and alleviation of his suffering" (no. 4, emphasis in the original).
4. *The need to resist making a person's life contingent on its quality:* "it is not enough to reaffirm the general principle according to which the value of a man's life cannot be made subordinate to any judgment of its quality . . . it is necessary to promote the *taking of positive actions* as a stand against pressures to withdraw hydration and nutrition as a way to put an end to the lives of these patients" (no. 6; emphasis in original).
5. *The principle of solidarity:* "It is necessary, above all, *to support those families* who have had one of their loved ones struck down by this terrible clinical condition" (no. 6; emphasis added).

## 3.3.2 Some Comments on the Address

Although the immediate context for the Pope's remarks was a conference on the vegetative state, similar ethical questions arise in the case of other patients, e.g. those suffering from advanced dementia, severe stroke, advanced metastases, quadruple amputees, etc. Hence, the Pope's speech is of wide relevance to Catholic healthcare professionals. In addressing these issues in the form of an allocution to a gathering of healthcare professionals, John Paul II followed the example of his recent predecessors who used similar contexts in exercising their ordinary authority to speak on ethical issues.

John Paul II emphasizes that the providing of food and water even by artificial means is to be regarded *"in principle"* (emphasis added) " *'ordinary' and 'proportionate'* and as such morally obligatory." Thus although such provision is obligatory in principle, the Pope allows for those cases in which the provision of nutrition and hydration would not be appropriate, either because they would not be metabolized adequately, or because their mode of delivery would be gravely burdensome.

As the Australian Bishops have noted, the Pope's statement does not explore the question whether artificial feeding involves a medical act or treatment with respect to insertion and monitoring of the feeding tube. While the act of feeding a person is not itself a medical act, the insertion of a tube, monitoring of the tube and patient, and prescription of the substances to be provided, do involve a degree of medical and/or nursing expertise. To insert a feeding tube is a medical decision subject to the normal criteria for medical intervention (2004).

### 3.3.3 Some Negative Responses to this Address

Among the principal negative responses to John Paul II's Address were the following: it marked "a significant departure from the Roman Catholic bioethical tradition," (Shannon and Walter, 2004, 18–20; Hamel and Panicola 2004, 6–13; Tuohey, 2004); it was not in conformity with the 1980 Vatican *Declaration on Euthanasia (Iura et bona)* (Shannon and Walter, 2004), and was not so primarily because it imposes excessively severe burdens on the families of such persons (Deblois, 2004; Paris, 2005). Thus John Paris, S.J., claimed that the pope's talk ran counter to over 400 years of Church teaching, that it mandated use of excessively burdensome means, that it was probably written not by the pope but by Bishop Elio Sgreccia, who "represents the radical right-to-life segment of thinking" (Paris, 2005). Edward Sunshine found the Pope's address so utterly incompatible with the 1980 *Declaration on Euthanasia* and that it is "merely an assertion of ecclesiastical authority, with little grounding in reason," (Sunshine, 2004), and John Tuohey likened it to a faulty thesis proposal by a graduate student ignorant of traditional Catholic teaching (Tuohey, 2004). The principal criticisms of John Paul II's address by Catholic ethicists are that it is incompatible with traditional Catholic teaching and that it imposes excessive burdens on the families of the permanently unconscious. I will thus focus on these criticisms in my defense of the document.

## 3.4 Defense of John Paul II's Address

### 3.4.1 Compatibility of the March 20 Address with "Traditional Catholic Teaching"

Prior to his address in 2004 John Paul II had issued a major Encyclical explicitly concerned with life issues, *Evangelium vitae,* in 1995. In it he explicitly appealed to the 1980 *Declaration on Euthanasia (Iura et bona)* issued by the Congregation of the Doctrine of the Faith, declaring, "Euthanasia must be distinguished from the decision to forgo...medical procedures which no longer correspond to the real situation of the patient, either because they are by now disproportionate to any expected results or because they impose an excessive burden on the patient and his family.... To forgo extraordinary or disproportionate means is not the equivalent of suicide or euthanasia; it rather expresses acceptance of the human condition in the face of death" (1995, no. 65). Thus John Paul II obviously agrees that, whenever medical treatment or the provision of nutrition and hydration is withheld or withdrawn for legitimate reasons (futility, burdensomeness), this is not euthanasia. In his 2004 address he reaffirms the traditional teaching of the Church that only ordinary or proportionate means to sustain life are morally obligatory, and he uses the traditional criteria to determine whether means are ordinary or not: the ratio of benefit to burden and effectiveness in providing care for the patient. His teaching in no way requires that tubally assisted feeding and hydration be maintained at all costs, but only when the *benefits such assistance provides are present and no excessive burdens are imposed.* If in particular instances such feeding/hydration would not effectively preserve life or alleviate suffering it would lack its beneficial effect and would be futile.

In summary, the Pope's statement is an application of traditional Catholic teaching, and says neither that nutrition and hydration must always be given, nor that they are never to be given, to unresponsive and/or incompetent patients. Rather, the Pope affirms the presumption in favor of giving nutrition and hydration to all patients, even by artificial means, while recognizing that in particular cases this presumption gives way to the recognition that the provision of nutrition and hydration would be futile or unduly burdensome.

His statement is fully compatible with the *Ethical and Religious Directives for Catholic Health Care Facilities,* whose directive no. 58 declares: "There should be a presumption in favor of providing nutrition and hydration to all patients, *including patients who require medically assisted nutrition and hydration, as long as this is of sufficient benefit to outweigh the burdens involved to the patient"* (emphasis added). This directive, which occurs in Part 5, should be read in light of what the bishops have to say in their introduction to Part 5. There they declare:

> Some state conferences, individual bishops, and the NCCB Committee on Pro-Life Activities have addressed the moral issues concerning medically assisted hydration and nutrition . . . these statement agree that hydration and nutrition are not morally obligatory either when they bring no comfort to a person who is imminently dying or when they cannot be assimilated by a person's body.

Note that here the Bishops of the United States explicitly refer to the document prepared by the NCCB Committee on Pro-Life Activities and to documents of some state conferences. Of these the most developed is that of the Bishops of Pennsylvania. It will thus be suitable to see how John Paul II's Address relates to these documents.

Take first the document prepared by the NCCB Committee for Pro-Life Activities, National Conference of Catholic Bishops in 1992. That document challenged the claim, made by some theologians, that it was permissible to withhold/withdraw tubally assisted food and water from persons in the "vegetative" state because such persons are not able to pursue the "spiritual purposes" of life. The Bishops' Committee explicitly stated that it did not find these reasons "persuasive." Like John Paul II in his March 20 Address, it declared nutrition and hydration generally a form of "ordinary care," or at least ordinary means of preserving life, and that withholding/withdrawing such nutrition and hydration is a form of euthanasia by omission when the intent is to end life (1992).

As I noted earlier, John Paul II himself singled out this paper for praise in a talk to a group of U.S. bishops on their *ad limina* visit to the Vatican in 1998.

From this it is abundantly evident that the position taken by John Paul II in his March 20 address is completely compatible with traditional Catholic teaching as understood by this important committee of US bishops and, it can be added, the entire US Conference of Bishops *Ethical and Religious Directives.*

### 3.4.2 Does John Paul II Impose Excessively Grave Burdens on Families?

As noted, a major criticism of John Paul II's address is that he in effect imposes grave burdens of the families of those whose lives are to be sustained by providing them with food and water by tubal means even when they are not consciously aware of themselves or others, and traditionally Catholic teaching has recognized that life-sustaining measures are disproportionate or extraordinary if they impose excessive burdens on patients *or* their families.

But simply providing the permanently unconscious with food and water does not impose burdens on families or other care-givers, just as the *feeding* of those paralyzed from the neck down or suffering loss of all limbs does not impose excessive burdens on care-givers. The burden they carry is not caused by the *feeding* but rather by the continued living of those for whom they care. But this is a burden that must in justice be accepted by others. Would those opposing John Paul II claim that we should stop feeding the demented, the paralyzed, quadruple amputees, and individuals who are simply "not with it"? Withholding or withdrawing tubally provided food and water would not eliminate the burden of care-givers; only the *death* of those cared for would end the burden. Here observations made by Germain Grisez are very relevant. He noted that as the permanently unconscious person's loved ones witness what is done to provide food and other care they experience a great and undeniable burden. He then noted:

> Of course, this burden will be eliminated if food is withheld, but only because the comatose person will be eliminated. Thus, to decide not to feed a comatose person in order to end the burden and his or her loved ones experience is to choose to kill that person in order to end the miserable state in which he or she now lives (1989, p. 171).

### 3.5 Conclusion

Here I have shown, I believe, that John Paul II's March 20 address on the care that must be given to persons in the "vegetative" state is in no way incompatible with the "Catholic tradition," but that it is to the contrary, fully compatible with that tradition.

Moreover, John Paul II was keenly aware of the great hardship that families of PVS patients endure in caring for them. He thus outlined some of the important positive steps that may be taken to help these patients and their families, and ". . . to stand against pressures to withdraw hydration and nutrition as a way to put an end to the lives of these patients" (no. 6). John Paul II suggested the following concrete practical ways to help:

the creation of a network of awakening centers with special treatment and rehabilitation programs; financial support and home assistance for families when patients are moved back home at the end of intensive rehabilitation programs; the establishment of facilities which can accommodate those cases in which there is no family able to deal with the problem or to provide "breaks" for those families who are at risk of psychological and moral burnout (no. 6).

Such steps would demonstrate society's concern and love for these seriously impaired individuals. From a specifically Christian perspective, they would give powerful testimony to the faithfulness and selflessness of Christian love. They would provide evidence of the genuine and disinterested character of Christian love, which continues to be expressed even when those who receive it can show no appreciation—even when they are apparently totally unaware of this loving presence.

# References

American Academy of Neurology. (1989). Position of the American Academy of Neurology on certain aspects of the care and management of the persistent vegetative state patient. *Neurology, 39*, 125–126.

American Medical Association, Council on Scientific Affairs and Council on Ethical and Judicial Affairs. (1990). Persistent vegetative state and the decision to withdraw or withhold life support. *Journal of the American Medical Association, 263*, 426–430.

American Medical Association, Judicial Council. (1981). *Current opinions of the judicial council of the American medical association. Including the principles of medical ethics and rules of the judicial council.* Chicago: American Medical Association, p. 9, (par. 2.11).

Ashley, B., O.P., & O'Rourke, K., O.P. (2002). *Health care ethics.* (4th ed.). Washington, DC: Georgetown University Press.

Ashley, B., O.P., O'Rourke, K., O.P., & J. Dublois, C. S. J. (2006). *Health care ethics.* (5th ed.). Washington, D.C.: Georgetown University Press.

Australian Bishops. (2004). Briefing note on the obligation to provide nutrition and hydration 09-05-04, http://www.acbc.catholic.org.au/pdf/040903_briefing_note.pdf.

Committee for Pro-Life Activities, National Conference of Catholic Bishops. (1992). *Nutrition and hydration: Moral and pastoral reflections.* Washington, D.C.: United States Catholic Conference. Publication No. 516-X, p. 7, (Reprinted in *Origins: NC News Service 21*, 705–711).

Curran, C. E. (2005). *The moral theology of Pope John Paul II.* Washington, D.C.: Georgetown University Press.

Doerflinger, R. (2004). John Paul II on the "vegetative" state: An important papal speech. *Ethics & Medics, 29*, 2–4.

Deblois, Sister J. M. C. S. J. (2004, Spring). Prolonging life or interrupting dying? Opinions differ on artificial nutrition and hydration. *Aquinas Institute*, Newsletter, available at http://www.ai.edu.

Gelineau, Most Rev. L., Bishop of Providence, R. I. (1988). On removing nutrition and water from a comatose woman. In *Origins: NC News Service, 17*, 546–547.

Grisez, G. (1986). A Christian ethics of limiting medical treatment. In D. Lescoe & C. Liptak, (Eds.), *Pope John Paul II Lecture Series in Bioethics*, (Vol. 2, pp. 49–50). F. CT: Holy Apostles Seminary. Cromwell, CT: Pope John Paul II Bioethics Center.

Grisez, G. (1997). *Difficult moral questions.* Quincy, IL: Franciscan Press.

Grisez, G. (1989). Should nutrition and hydration be provided to permanently unconscious and other mentally disabled persons? *Issues in Law & Medicine, 5,* 165–179.

Hamel, R., & Panicola, M. (2004, April). Must we preserve life? *America, 19–26,* 6–13.

Hastings Center. (1987). *Guidelines on the termination of life-sustaining treatment and the care of the dying.* Briarcliff Manor, NY: The Hastings Center.

John Paul II, Pope. (1998, October 15). Building a culture of life. *Ad Limina* address to the bishops of California, Nevada and Hawaii. *Origins, 18*(4), 314–315.

May, W. E. (1977). *Human existence, medicine and ethics: Reflections on human life.* Chicago: Franciscan Herald Press.

May, W. E., Barry, R., O.P., Griese, O., Grisez, G., Johnstone, B., Marzen, C.Ss.R. et al. (1987). Feeding and hydrating the permanently unconscious and other vulnerable persons. *Issues in Law & Medicine, 3.* 203–217.

McHugh, J., & Bishop of Camden, N. J. (1989). Artificially assisted nutrition and hydration. *Origins: NC News Service, 19,* 314–316.

New Jersey State Catholic Conference. (1987). "Friend-of-the-court brief to the New Jersey supreme court": Providing food and fluids to severely brain damaged patients. *Origins: NC News Service, 16,* 542–553.

O'Rourke, K., O. P. (1986). The A.M.A. statement on tube-feeding: an ethical analysis. *America, 155,* 321–333.

O'Rourke, K., O.P. (1988). Evolution of church teaching on prolonging life. *Health Progress, 59,* 28–35.

O'Rourke, K., O.P. (1989). Should nutrition and hydration be provided to permanently unconscious and other mentally disabled persons? *Issues in Law & Medicine, 5,* 181–96.

O'Rourke, K., O.P. (1989). Open letter to bishop mchugh: father kevin o'rourke on hydration and nutrition. *Origins: NC News Service, 19,* 351–52.

O'Rourke, K., O.P. (1999). On the care of "vegetative" patients. *Ethics & Medics 24.4,* 3–4 and *24.5,* 1–2.

O'Rourke, K., O.P. & Ashley, B., O.P. (1997). *Health care ethics* (4th ed.). Washington, DC: Georgetown University Press.

O'Rourke, K., O.P. & Norris, P., O.P. (2001). Care of PVS patients: catholic opinion in the United States. *Linacre Quarterly, 68.3,* 201–17.

Paris, J. S. J. (2005). No moral sense. in an interview with brian braiker of *newsweek* and available at: http://www.msnbc.msn.com/id/7276850/site/newsweek/.

Pennsylvania Conference of Catholic Bishops (1992). Nutrition and hydration: moral considerations. *Origins: NC News Service, 21,* 542–553.

Pius XII, Pope. (1958). The prolongation of life: Allocution to the international congress of anesthesiologists of November 24, 1957. *The Pope Speaks, 4,* 96.

Pontifical Academy of Sciences, Report on the Artificial Prolongation of Life. (1985). Origins: NC news service. Reprinted In A. Moraczewski, O.P., & R. E. Smith, (Eds.), *Conserving human life.* (p. 306). Braintree, MA: Pope John XXIII Medical-Moral Research and Educational Center,.

President's Commission for the Study of Ethical Problems in Medicine and Biomedical and Behavioral Research. (1983). *Deciding to Forego Life-Sustaining Treatment: Ethical, Medical, and Legal Issues in Treatment Decisions.* Washington, DC: U. S. Government Printing Office.

Shannon, T., & Walter, J. (2004). Implications of the papal allocution on feeding tubes. *Hastings Center Report, 34,* 18–20.

Shewmon, D. A. (1997). Recovery from "Brain Death": A neurologist's apologia. *The Linacre Quarterly, 64,* 59–60.

Sunshine, E.R. (2004). Truncating catholic tradition. *National Catholic Reporter*, at http://www.natcath.com/NCR_Onlinearchives2/2005b/040805k.php.

Texas Conference of Catholic Bishops (1990). On withdrawing artificial nutrition and hydration. *Origins: NC News* Service, 20, 53–55. It should be noted that 2 of the 18 members of the Texas Conference of Catholic Bishops refused to sign this statement.

Tuohey, J. (2004). The pope on PVS: Does JPII's statement make the grade? *Commonweal, 131*, 10–13.

Walter, J., & Shannon, T. (1988). The PVS patient and the forgoing/withdrawing of medical nutrition and hydration. *Theological Studies, 49*, 623–647.

# Chapter 4
# Food and Fluids: Human Law, Human Rights and Human Interests

Jacqueline Laing

## 4.1 Introduction

Academic discussion about nutrition and hydration tends to concentrate on conceptual matters intrinsic to the ethics of removing food and fluids in individual cases. It is, for example, undoubtedly important to distinguish between vitalistic and utilitarian excesses in understanding the rights and wrongs of withdrawing food and fluids delivered by tube or by spoon from mentally incapable patients. Vitalism wrongly insists that *all* must be done to save the life of the incapacitated patient irrespective of the legitimate wishes of the patient, and the cost, effectiveness and physical burden on the patient of the intervention in question. Utilitarian accounts wrongly sacrifice the principle of the inherent dignity of every human being however disabled to a "quality of life" principle insisting that some people lack personhood or have disabilities that suggest that their very lives (as distinct from their treatment) should be regarded as undignified, futile or even over-burdensome.

In the context of changing positive law, however, it is important to understand the considerable financial, scientific and medical interests there are in controlling death. These interests need not be illicit in themselves. The interests of hospital and state efficiency, freedom from unnecessary compensation claims, scientific research and increased supplies of organs for transplant are not in themselves wrongful. When understood in the context of law that invites bureaucratised homicide and serious mutilation of the non-consenting or ill-informed vulnerable, these interests introduce new extrinsic concerns. There is every reason to believe that a proper analysis of this ethico-legal terrain demands a comprehensive inquiry into wider matters sometimes wrongly rejected as consequentialist. Failure to identify these broader interests and their moral limits might well lead one to a conceptual failure to see the wood for the trees.

England and Wales has seen radical alteration of the law of homicide and assault. The *Mental Capacity Act 2005* (which comes into force in 2007) will soon

J. Laing
Human Rights and Social Justice Research Institute, London Metropolitan University
Email: j.laing@londonmet.ac.uk

C. Tollefsen (ed.), *Artificial Nutrition and Hydration:*
*The New Catholic Debate*, 77–100. © Springer 2008

govern the removal of "treatment" which, after *Airedale NHS Trust v Bland* [1993] AC 789, includes food and fluids delivered by tube, and in certain cases also, by spoon. It does so by introducing binding advance decisions, attorneys empowered to make certain treatment decisions on behalf of the patient and a new version of the Court of Protection which will replace the jurisdiction of the ordinary courts. It also consolidates and extends recent case law permitting sterilisation and abortion on those considered incapacitated. It permits non-therapeutic research on non-consenting mentally incompetent adults. By recognising the binding nature of the advance decision, it sets up the conceptual apparatus for introduction of routine administration of the lethal injection. It introduces the notion of an attorney newly empowered to make certain "treatment" decisions on behalf of the patient. Given the abuse and homicide it arguably invites, it is possible to see the legislation as a responsibility-shifting exercise designed to foster new socially useful but fundamentally unjustly won ends.

The UK has also passed the *Human Rights Act 1998* which introduces into English law the *Convention for the Protection of Human Rights and Fundamental Freedoms* (hereafter *European Convention on Human Rights*, the Convention or ECHR). Whatever one's reservations about the actual application of the Convention in particular cases and its role in European domestic legislatures, the *Human Rights Act 1998* is now undoubtedly a part of the positive law of the UK. The conceptual apparatus of the Convention is far from antithetical to a genuinely natural law bio-ethic. The language of the Convention recognises, at least on its surface, the intrinsic dignity of human beings and, properly understood, permits a genuinely natural law ethic of dying. It is therefore instructive to visit the positive law of England and Wales on withholding and withdrawing food and fluids, examining it against the demands of the *European Convention on Human Rights* and against the background of financial and scientific interests in controlling death, dying and the human body itself. In what follows, tensions between the 2005 Act and the *European Convention on Human Rights* are examined. I argue that theoreticians and lay folk alike are being persuaded of the need for this legislative reform on the basis of unsustainable readings of personal autonomy and the social good, on an improper understanding of the ethical principles governing human intervention in death and dying, and in ignorance of the substantial financial and scientific interests behind the legislation. Far from promoting autonomy and the social good, the legislation undermines human rights and threatens human dignity.

## 4.2 The Mental Capacity Act 2005

The *Mental Capacity Act 2005* (hereafter, the 2005 Act) has significant implications for mentally incapacitated patients in England and Wales.[1] It constitutes the culmination of efforts by successive governments proceeding from the Law Commission *Draft Bill on Mental Incapacity 1995*, to enact legislation in respect of the care and treatment of the mentally incapacitated. Read in the light of existing case law, certain sections of the *2005 Mental Capacity Act* have profound consequences. Most notably they give a catalogue of new actors power to withhold and withdraw "treatment" including artificial nutrition and hydration from patients who, it should be highlighted, may not be dying. These new decision-makers include donees under lasting powers of attorney (attorneys) and those purporting to bear the advance decisions of mentally incapacitated patients. In addition, wide-ranging powers are established in respect of a virtually unrecognisable Court of Protection now empowered to make life and death decisions governing removal of "treatment" as well as decisions to perform research on, remove tissue from, sterilise and abort the young of mentally incapacitated patients.

The *Mental Capacity Act 2005* needs to be read in conjunction with other legislation that has appeared recently. The *Human Tissue Act 2004* (which came into force in 2006) permits *inter alia* use of tissue from non-consenting patients. The *Medicines for Human Use (Clinical Trials) Regulations 2004* (S.I. 2004/1031) allows for clinical drug trials on non-consenting patients on the authority of novel representatives. The *Mental Capacity Act 2005* also expressly permits non-therapeutic research on the non-consenting on the authority of novel third parties. As we shall see, these proposals exist against an intellectual background that can be described as broadly utilitarian. In successive volumes of *The Lancet*, senior medico-legal figures (Hoffenberg et al., 1997, pp. 1320–1321) representing the International Forum for Transplant Ethics make the case for removal of organs from non-consenting patients in permanent vegetative state for use in transplantation. They also recommend societal opt-out organ "donation" as a way of increasing the stock of organs available for transplant (Kennedy et al., 1998, pp. 1650–1652). For non-utilitarian bio-ethicists these suggestions might highlight the aims, driving interests and moral limits of the legislation.

It is also worth remembering too that some twenty years earlier in 1984, at the 5th Biennial Conference of the World Federation of Right to Die Societies held in Nice, Australian bioethicist Dr. Helga Kuhse suggested a strategy for the implementation of euthanasia by lethal injection: "If we can get people to accept the removal of all treatment and care—especially the removal of food and fluids—they will see what a painful way this is to die and then, in the patient's best interest, they will accept the lethal injection" (Marker, 1993, pp. 94, 267).

It is widely argued that this law reform is progressive, fosters patient autonomy and clears the way for necessary scientific research. An alternative, more realistic reading is that these radical alterations in the law of assault and homicide create contradictory and unworkable obligations for health professionals and fundamentally compromise the human rights and bodily integrity of the vulnerable.

## 4.3 The Background to the 2005 Act

The 2005 Act allows new agents to require doctors to withdraw or withhold treatment from mentally incapacitated patients. Ever since the controversial and highly criticised House of Lords decision in *Airedale NHS Trust v Bland* (1993) AC 789, treatment has included tube feeding and even feeding by hand in cases where this is possible. So what the proposed legislation logically authorises is the removal of food and fluids with consequent dehydration to death of patients.

When *Bland* was decided, the case attracted much criticism not least because three out of five Law Lords stated that the aim of stopping feeding was to bring about Tony Bland's death. *Bland* was understood, by both supporters and critics of the decision, to mark a *volte face* in English law. Well-known euthanasia advocates like Peter Singer (1994, p. 1), for example, noted that the case marked the collapse of the Judeo-Christian principle of the inviolability or sanctity of human life. Critics regarded the apparent rationale behind the decision defective (for example, Finnis, 1993, p. 329), and argued that the doctrine of the sanctity of life had been "misrepresented, misunderstood and mistakenly rejected" (Keown, 1997, p. 481). The majority's reasoning involved three important propositions. The first was that tube feeding was "treatment" not ordinary care (Lord Keith, *Bland*, p. 858). For the first time tube feeding was regarded as treatment. The second and most important proposition in the majority's reasoning was that Tony Bland had no "best interest" because he had *no meaningful life* (Lord Mustill, *Bland*, p. 897). The third proposition was that while it would have been unlawful to kill Tony Bland with a lethal injection, removal of his feeding tube would constitute a permissible omission (Lord Goff, *Bland*, p. 868).

### 4.3.1 *"Treatment" or Ordinary Care?*

So in the UK now, *Bland* has come to stand for the proposition that tube feeding is not ordinary care but "treatment" which, in certain circumstances, may be withdrawn even from people who would not otherwise die.[2] The result of withdrawing tube feeding from a patient and then refraining from feeding the patient by hand (this is often possible even with patients diagnosed in a persistent vegetative state, or PVS) is that the patient dies some days later of hunger and thirst. It is well known that feeding by tube is a simple matter. It makes life easier for nurses and other health professionals who might otherwise have to spend some hours feeding by hand. The tube can either be placed in the abdomen or inserted by a capable patient himself or herself through the nose. Its point is to ease feeding, a natural function of the body. It is not costly. It is simple basic care, a non-technical extension of the every-day activity of feeding by hand. Above all, it is not an attempt to stabilize, treat or cure a patient, as is something like ventilation.

That tube feeding should be regarded as effective ordinary care is emphasized by Keith Andrews of the Royal Hospital for Neurodisability in South London. Keith Andrews is particularly well placed to comment. He was a witness in *Bland* and it was he who later documented 17 out of 40 misdiagnoses of PVS some three years after *Bland* was decided (Andrews et al., 1996, pp. 13–16).[3] He has been reported as saying that: "the only reason that tube feeding has been identified as 'treatment' is so that it can be withdrawn...I would argue that tube feeding is extremely effective since it achieves all the things we intend it to. What is really being argued is whether the patient's life is futile—hence the need to find some way of ending that life" (1995, p. 1437). His analysis highlights a most important feature of the *Bland* decision. After the enactment of the 2005 Act this part of the *Bland* majority judgement has profound implications.

A fact worth mentioning about the aftermath of the *Bland* case is that nearly a decade after the initial injury another survivor of the Hillsborough disaster woke up. Like Bland, he was diagnosed as in PVS. Andrew Devine was apparently in a "permanent" vegetative state for eight years before communicating with his parents.[4] Stanley and Hilary Devine, the parents of Andrew, had never sought to prevent their son's being fed. In 1996, the solicitor for the Devines reported a massive improvement and expressed the family's wish for privacy. In fact, there are a number of examples of patients awaking[5] from a "permanent" vegetative state.[6] So frequently have diagnoses of "permanent vegetative" state been falsified by the patient's subsequent recovery or further scientific revelations, that the very terminology used has been altered. "Permanent" is now "persistent" vegetative state. The word "vegetative" is still common parlance despite the pejorative connotations. I adopt the prevailing terminology to avoid misunderstanding and despite its dehumanising overtones. Once a person is regarded a vegetable or an animal, it become less difficult to permit the bringing about of his death.

## 4.3.2 *"Worthless Lives" and "No Best Interests"*

Recent first instance cases have taken the *Bland* decision as authority for the idea that some people have no meaningful lives and therefore no "best interests." Since *Bland* the courts have been at liberty to make this determination before withdrawing tube feeding. In *Re D (Adult: Medical Treatment)* [1998] 1 FLR 411 the patient was able to respond to ice water, was able to track moving objects and evinced a "menace" response. It was held, applying *Bland*, that notwithstanding the fact that the criteria for PVS were not fulfilled, the patient showed no evidence of a meaningful life and that it was not in D's interest to be "kept alive."

Again in *Re H (A patient)* [1998] 2 FLR 36, the patient could focus on an object and could be aroused by clapping. There was evidence of visual tracking as well. It was held that H was in PVS and that cessation of "treatment" was in her best interests. What is particularly disturbing is that the patients involved were not dying. They died finally of dehydration once tube feeding was withdrawn.[7]

This new approach to the mentally incapacitated derives from certain majority judgements in *Bland*. Lord Hoffmann described Tony Bland thus: "His body is alive, but he has no life in the sense that even the most pitifully handicapped but conscious human being has a life". He went on to describe Tony Bland's existence as a humiliation. He was, he said, "grotesquely alive" (Lord Hoffmann, *Bland* p. 863). Lord Keith referred to Tony Bland's "existence in a vegetative state with no prospect of recovery [as considered by responsible medical opinion] as not being a benefit..."

(Lord Keith, *Bland*, p. 858–859) Lord Mustill asserted that Tony Bland "had no best interests of any kind" (Lord Mustill, *Bland*, p. 897).

At the time of the decision it was pointed out that these kinds of statement suggested a new drive, one which sought to determine whether a person's life is of a sufficiently high *quality* to warrant the protection of the law. This idea of a "worthless life" and the companion question "whether it is in the best interests of a patient to survive" is a new one and arguably runs contrary to the criminal law as traditionally understood. It has, until recently, been an assumption of the law that all human beings share the same fundamental worth simply in virtue of their humanity irrespective of their physical or mental abilities or disabilities. The law has steadfastly refused to discriminate between those thought to have worthwhile lives and those pronounced worthless.

A central problem with the notion of a "worthless life" (for the purposes of permitting some rather than other intentional homicides), is that the notion, upon analysis, is fraught with difficulty.[8] Above all the concept involves *unjust discrimination* against the severely disabled precisely on the basis of the severity of their disability. Furthermore, the notion of a "worthless life" is highly subjective and fraught with arbitrariness. This, in itself, invites abuse. Health professionals and observers are faced with laws that are neither stated nor promulgated. They cannot know in advance whether they are obeying the law or are in breach of it. Without clear and public criteria for deciding these matters, the law itself becomes an instrument of injustice operating on an entirely unpredictable basis. I have addressed the inherent discrimination against the disabled implicit in the concept of the "worthless life." The concept has a certain verisimilitude to the Nazi notion of the "*lebensunwerten Lebens*" in any case. The principal objection is that we all suffer and are vulnerable at one time or another. Attempts to stipulate criteria like "rationality, self-consciousness and autonomy" as the necessary conceptual test, have the unhappy and counterintuitive consequence of suggesting that the sleeping, the drunk and the unconscious lack the technical status of "personhood" (Laing, 1996, pp. 196–225). Even the judge and professor must sleep and countless people find themselves, at some point in their lives, unconscious. Who is to say that the non-responsive patient (or those otherwise incapacitated) is living the life of the non-person, or the life unworthy of life? At the very least, if there is to be a general concept of a "worthless life" or "non-person" the criteria for the application of the notion had better be clear and identifiable. No such criteria were made plain in *Bland*. As we shall see in what follows, the technical concepts of "lives unworthy of life" and non-personhood are arguably ones that are antithetical to the concerns for equal dignity, human life and just treatment contained in the European Convention on Human Rights, and other international instruments.

### 4.3.3 Intentional Killing by Omission

The third revolutionary aspect of the *Bland* decision was its approach to intentional killing by omission. Traditionally, intentional killing by omission was prohibited. The authors of a standard and authoritative British textbook of criminal law, Smith and Hogan, now describe the decision in *Bland* thus: "There was no doubt about the intention to kill. The object of the exercise was to terminate B's life. It was accepted that to kill by administering lethal injection or any similar act would be murder; but what was proposed was held to be not an act but an omission" (Smith and Hogan, 1999, p. 50).

It is indeed a long-established principle of the common law that there is no duty to save a person from death. If you or I see someone drowning in the sea, there is no obligation to dive in and save the victim. It is this idea that the majority in *Bland* relied upon in granting the declaration to withdraw tube feeding. Since Tony Bland had no right to "treatment" (here, tube feeding), he was not being deprived of anything to which he had a right when it was removed, with whatever purpose. This, at least, was the majority's rationale.

But, as was pointed out at the time (Finnis, 1995, p. 329; Keown, 1997, p. 481), intentional killing by omission is still murder provided the intention is there. If I intend to kill my baby at home by omitting to feed it, it is the fact that I intend to kill that is important in determining whether this omission should be regarded as murder.[9] The fact that the method used to kill is an omission will not save me from a murder conviction if the intention can be proved. It might be difficult to prove intention, as it is in many other kinds of case, but evidential problems are not substantial ones. It is a long established principle that murder can be committed by intentional omission as well as by intentional act.[10] It is also well-known that manslaughter can be committed by omission where there is an assumption of care of the victim or where there is a special relationship or a special duty to act created by statute or contract or public office. There can be no doubt at all that the doctor-patient relationship involves, in the most intimate way, this duty of care.

So the majority's decision to permit intentional killing by omission by health professionals in circumstances where the patient was not dying marked a major break with the English criminal law. If *Bland* indicated willingness by the English courts to break with existing English law, the *Mental Capacity Act 2005* goes much further. It permits various third parties, attorneys and those claiming to know the advance decisions of the patient, to authorise what the courts alone after *Bland* were authorised to order.

## 4.4 Human Rights and the 2005 Act

The notion of equal dignity informs Article 2 of the *European Convention for the Protection of Human Rights and Fundamental Freedoms* which provides that:

> Everyone's right to life shall be protected by law. No-one shall be deprived of his life intentionally save in the execution of the sentence of a court following his conviction of a crime for which this penalty is provided by law.

Given that dehydration is a particularly nasty way to die the principle is also borne out by Article 3 which states that: "No one shall be subjected to . . . inhuman or degrading treatment or punishment." Article 8 states that: "Everyone has the right to respect for his private and family life, his home and his correspondence." It is also made explicit in Article 14 which stipulates that:

> The enjoyment of the rights and freedoms set forth in this Convention shall be secured *without discrimination* on any ground such as sex, race, colour, language, religion, political or other opinion, national or social origin, association with a national minority, property, birth or other status.

The fact that a person is disabled, even severely disabled, is no grounds to discriminate against his right to life and to freedom from inhuman and degrading treatment. Notwithstanding these articulations recent human rights cases suggest the position in relation to artificial nutrition and hydration is far from settled.

On 30 July 2004, shortly after the *Mental Capacity Bill* received its first reading, the High Court handed down an important judgment analysing the General Medical Council's guidelines on withdrawing and withholding food and fluids. In that case Leslie Burke, a man with a progressive neurological condition, perceiving the combined effects of the Bland case and the effects of the GMC guidance which permitted the removal of food and fluids on quality-of-life grounds, sought a declaration that the guidance failed to protect against human rights abuses. Mr Justice Munby found that the GMC guidance was indeed defective because it allowed artificial food and fluids to be withdrawn from patients in circumstances that failed to protect against breaches of Article 2 (right to life), Article 3 (freedom from inhuman treatment), Article 8 (right to family and private life) and Article 14 (non-discrimination) of the European Convention.

No sooner had judgement been handed down, but the government announced its intention to appeal. The Court of Appeal duly overturned the decision of the High Court and, finally, on appeal to the European Court of Human Rights, the Court considered that Mr Burke had failed to establish that UK law was such that he faced a real or imminent risk that tube-feeding would be withdrawn in circumstances that implied a painful death by thirst. The Court stated that it was satisfied that the presumption of UK law was in favour of "prolonging" life wherever possible. The Strasbourg Court agreed with the Court of Appeal that the GMC Guidelines which Leslie Burke sought to challenge simply set out good practice for doctors and did not alter the law. They approved the Court of Appeal's judgment and confirmed that if a doctor withdrew tube-feeding from a *competent* patient who desired tube-feeding to continue then it would be murder. Where a patient was *incompetent*, however, then as a general rule they considered tube-feeding should continue for as long as it prolonged life. There were, however, circumstances where a doctor might find that ANH in fact hastened death and thus it was impossible to lay down any absolute rule as to what the best interests of a patient would require.

The unwillingess of the Strasbourg Court to enter into a debate about UK law was perhaps only to be expected given that the newly enacted *Mental Capacity Act 2005*, with its then unpublished *Code of Practice*, had yet to come into force. Had the challenge been successful the *Burke* Case would have pre-empted the 2005 Act and trumped parasitic instruments still in a draft stage.

There was considerable opposition to the *Mental Capacity Bill,* not merely to the worrying dehydration questions raised by the Bill but also to other matters. Novel third parties (such as attorneys, those claiming to have legally binding advance directives refusing treatment, and, in the early stages of the Bill's passage too, court appointed deputies) were authorised to require doctors on pain of an assault charge, to remove and withhold "treatment" (which after *Bland* includes ANH and in certain cases spoon feeding too). Not only this but controversial procedures like non-voluntary sterilisation and non-voluntary abortion (then questionably permitted but only on a High Court order *Re F (Mental Patient: Sterilisation)* [1990] 2 AC 1; *Re SG* [1991] 2 FLR 329 respectively) were, at that stage, potentially in the hands of these newly empowered agents.

A profoundly different Court of Protection was emerging, one that no longer merely dealt with the financial welfare of the incompetent but overseeing his very medical treatment, life and death. Importantly, the Bill allowed non-therapeutic research to be performed on certain mentally incapacitated patients without their consent. It abolished the High Court's jurisdiction to hear applications on the above-mentioned matters with the substitution of (and even then only in certain cases) the Court of Protection, an institution that then afforded very little of the transparency, requirement of representation, ordinary appeal and procedural form demanded by a genuine court. Indeed the 23rd Report of the Joint Committee on Human Rights confirmed these and a number of other concerns. The Bill, it argued, would also involve arbitrary deprivations of liberty occasioned by insufficient procedural safeguards as outlined in *Winterwerp v Netherlands* (1979) 2 EHRR 387 and *HL v United Kingdom* (Application No 45508/99) (unreported) 5 October 2004, and use of easily alterable Codes of Practice to specify matters that affect the law of homicide and assault thus suggesting an absence of procedural safeguards against abuse of fundamental human rights.

Although welcomed by numerous parties such as the Making Decisions Alliance, the Law Society, as well as the Voluntary Euthanasia Society, the Bill was opposed root and branch by other disability rights groups such as Disability Awareness in Action, People First, the British Council of Disabled People, the Coalition of Organizations of Disabled Peoples and I Decide. Numerous religiously affiliated organisations such the Evangelical Alliance, CARE, the Christian Medical Fellowship, the Lawyers Christian Fellowship, anti-euthanasia organisations such as ALERT, the Society for the Protection of Unborn Children, the British Section of the World Federation of Doctors Who Respect Human Life and anti-eugenics organisations also opposed the Bill (Laing, 2005b, pp. 137–143). Strangely, the Catholic Bishops Conference of England and Wales did not oppose the legislation root and branch but sought amendments only,[11] making no mention of the potential for numerous human rights abuse of the kind outlined.

## 4.5 The Apparatus: Advance Decisions, Attorneys, Deputies and a New Court of Protection

The Act permits for the first time in English law a variety of new agents to bind a doctor on pain of an assault charge, to remove "treatment" from the mentally incapacitated. Binding advance decisions, donees under lasting powers of attorney and a new Court of Protection are among the novel third parties empowered to require this state of affairs.

### 4.5.1 Advance Decisions

The trouble with binding advance decisions is precisely that they are made in advance either verbally or in writing. This refusal of treatment might occur when a person is not suffering from a particular condition, is not being offered any particular treatment, and has no idea what the condition requires or how he or she would feel in this particular situation. The patient may be speculating years in advance of the treatment. The advance decision legally *binds* the doctor. In the absence of complex inquiries into whether the statement constitutes a patient's up-to-date wishes, the decision determines the patient's fate and health professionals are legally indemnified for their lethal actions and omissions. An advance directive might be entirely nonsensical, medically speaking. The doctor would be bound by this document, often made long ago and in ignorance of the circumstances in which the patient finds himself.

It should be remembered that after *Bland* "treatment" means tube feeding (and in certain cases too, spoon-feeding). Many people simply do not know that an advance refusal of treatment may mean death by dehydration *at a time triggered and determined by the health team* who would be legally indemnified against homicide by virtue of the patient's advance decision. The Code specifies that a person *may* help himself to legal and medical advice in making his advance decision but does not require him to do so. For those who do not understand the law, the consequences are likely to be grave. Section 26(3) states that:

> A person does not incur liability for the consequences of withholding or withdrawing a treatment from P, if at the time, he reasonably believes that advance decision exists which is valid and applicable to the treatment (2007 Code, Clause 26 (3)).

What this means is that the health service is legally indemnified against prosecution, claims of negligence or disciplinary proceedings once the advance decision is triggered. The advance decision-maker may make a decision in ignorance of the implications of *Bland*, or of what cures will become available, or of how he will feel at the time in question. He would be bound, in the absence of complex investigations into the decision itself, by that refusal of life-sustaining treatment. The only safeguard, if such it can be called, is that an advance decision to refuse ANH must be in writing.

The binding advance decision envisaged by the 2005 Act reverses important presumptions in favour of saving life with the threat of litigation. This seriously undermines medical professionalism and the core ethic of the medical profession. The practical implication of being able to prosecute and sue a doctor for administering "treatment" in the face of an ill-informed advance decision, where it would have the effect of saving the patient from death or chronic disability, is that medical teams will be highly unlikely to undertake further investigation in emergency situations (as are often undertaken in respect of suspect wills with the benefit of time and cool consideration) into whether a person's refusal of treatment was properly informed and genuine rather than fraudulent, unconscionably obtained, or undertaken in ignorance. Since they would be indemnified against liability by triggering an advance decision, the medical system would be loaded against saving life even by relatively simple means. This creates a climate fundamentally hostile to the practice of medicine.[12]

If perchance the doctor *were* to act on his own initiative to give the patient the best, most up-to-date treatment thereby attempting to save the patient from long-term disability, he would be open to a charge of assault. And if he were to withhold treatment from his patient in accordance with the directive, he might yet find himself faced with a suffering, disabled patient properly anxious for damages for his disability. After all, how was the patient to know that his advance decision was going to leave him chronically disabled? The doctor, as we have seen, would be indemnified by section 26(3) against liability for the long-term disability occasioned by the existence of the directive. Thus, the binding advance decision threatens the vulnerable by undermining the position of patients who are left permanently or chronically disabled by the failure to receive treatment that they might otherwise have received were it not for the ill-informed advance directive. It also acts as the preventative for cure. Because bad clinical practice becomes binding on the doctor, the patient would have no straightforward recourse to the law of negligence.

This possibility in turn supplies a further cause for concern. The effect of the advance decision is to *shift responsibility* for significant clinical decisions to the decision-maker himself. There are substantial conflicts of interest involved in the business of shifting legal responsibility for lethal decisions since the health service and medical professionals themselves are no longer bound by a duty to act in the best clinical interests of the patient once the decision was triggered. These are undoubtedly pressures associated with compensation claims for bad treatment. Secondly, there are pressures on beds and resources. The Western world has a growing costly, non-productive ageing population thanks, in part, to its unwillingness to reproduce. Accordingly, top-down bureaucratic pressures to clear beds and increase hospital efficiency are bound to constitute an operating factor in the determination of whether a decision to withdraw "treatment" should be pursued. Thirdly, as we shall see, there are numerous other scientific and medical interests in controlling death. These possibilities might be deprecated as alarmist and over-pessimistic about medical good-will, but a brief consideration of the bureaucratic, financial, medical and research pressures on health professionals working in a fundamentally altered moral climate must give us pause.

In reality, binding advance decisions, made long in advance of known situations, introduce all manner of conflict and contradiction for the well-meaning health professional. There remains existing domestic law prohibiting assisting in suicide (*Suicide Act 1961*), recently confirmed as compatible with the Convention in *Pretty v. United Kingdom* [2002] 2 FLR 45. Accordingly, if the doctor permanently removed tube feeding from a patient whose *aim* it was to die by starvation, there would be the further possibility that he would be faced with a charge of assisting suicide. So he would be in breach of the law prohibiting assisting suicide if he acted on the advance directive. Damned if he did and damned if he did not, there would, in other words, be a straightforward conflict of obligations: i.e. to remove the food and fluids and participate in the patient's suicidal intent and to refrain from removing food and fluids on pain of an assault charge (Laing, 1990, pp. 106–116). Such would be the logical effect of enacting legislation allowing advance directives given the *Bland* decision. That this is not as fanciful as it might first appear is supported by the new Kelly Taylor[13] test case in which a patient wishes to require doctors to sedate her and then dehydrate her to death in a bid to hasten her death by a year. Cynics will doubtless see the case as one that is designed to further the aims of advocates of state sanctioned medical killing. Once it is admitted that intentional killing may be performed by omission a year in advance of natural death at the behest of the patient, the question of the lethal injection is the next logical step.

Much could be said about the contradictions raised by this new legislation. It is perhaps these grave deficiencies that prompted the warnings of the 23rd Report of the Joint Committee on Human Rights highlighting the failure of the legislation to supply adequate safeguards against Articles 2, 3 and 8 incompatibilities. Further, the fact that it is the mentally incapacitated as a class that are thought ripe for these and other kinds of intervention, highlights the Article 14 discrimination inherent in this and related legislation. For our purposes what remains of importance are the financial, medical and research interests that underpin the legislation and, in this context, the responsibility shifting exercise envisaged by section 26(3).

### 4.5.2 Attorneys

If the binding advance decision undermines personal autonomy in unexpected ways, the attorney deciding for the mentally incapacitated explodes it altogether. As I have suggested already, it is this responsibility shifting aspect of the Act that is perhaps its most dangerous feature. Substituted consent is not an expression of the personal autonomy of the patient. On the contrary, it is an expression of the autonomy of the attorney.

The 2005 Act introduces the concept of the "lasting power of attorney." We are familiar with the need for enduring (or durable) powers of attorney allowing certain people to deal with the property and *financial* affairs of the incompetent patient. The Act extends the ambit of existing powers to include medical and indeed life-and-death decision-making. Section 11(8) of the Act states that a lasting power of attorney extends to refusing consent to the carrying out or continuation of *life-sustaining treatment* where the lasting power of attorney contains express provision to that effect (section 11(7)(c)). Once again the accepted definition of "treatment" in *Bland* logically implies that the donee of a lasting power of attorney has the power to decide whether a patient should be dehydrated to death by the refusal of tube-feeding qua "treatment". It should be remembered that on occasion patients who are not dying at all will need ANH. They may be sedated or in a coma or unconscious. To allow a third party to substitute his consent for that of the patient is to invite abuse. This need not be malicious, though it may be. It may be simply undertaken in ignorance on the advice of those with a conflict of interest.

Again, it should be remembered that after *Bland* "treatment" means tube feeding (and in certain cases too, spoon-feeding). Many people simply do not know that an advance refusal of treatment may mean death by dehydration at a time triggered and determined by the attorney. Further there is no requirement that people filling in these new powers of attorney forms be advised of the legal implications of their decision. This again invites appalling abuse and shifts responsibility for profound decisions to those who will often be the least informed.

Now, where the lasting power of attorney authorises the attorney to make decisions about the patient's personal welfare, the authority by section 11(7) (c) extends to giving or refusing consent to the carrying out or continuation of a treatment by a person providing health care for the patient. Certain recent cases are authority for the proposition that non-voluntary sterilisation and non-voluntary abortion (*Re F (Mental Patient: Sterilisation)* [1990] 2 AC 1; *Re SG* [1991] 2 FLR 329 respectively) are "treatment" that can be in a mentally incapacitated, often mentally disabled, patient's "best interests." So the logical upshot is that the attorney can now authorise these profoundly questionable procedures and indeed the law sets up the apparatus for this new regime. The most worrying aspect of this novel responsibility-shifting initiative is that enormous burdens are placed upon the attorney as well as permitting hitherto unknown power to substitute his consent for that of the patient. The attorney bears a duty of care, good faith, and confidentiality as well as duties to comply with the directions of the Court of Protection. Thus, if the patient is left chronically disabled for refusal or authorisation of treatment, it will no longer be the health service to whom the patient must turn for legal redress, it will be his own attorney, often a loved one placed in this invidious position. There is, of course, the possibility that an attorney will be acting in bad faith.

Abuse is prohibited by the Act and the Code. Of greater concern, is precisely that the new attorney will often be acting on the advice of health professionals. When a health professional can be sued for his actions there is far greater likelihood that he will act in the best interests of the patient. Given that health professionals no longer bear primary responsibility for authorising controversial procedures like refusal of treatment (including food and fluids), abortion, sterilisation, research and other procedures, the question of the soundness of the advice being given will be pivotal. In short, once again far from promoting personal autonomy, the device of the substitute decision-maker often acting on bad medical advice (possibly driven by a conflict of interest) suggests the legislation invites Articles 2, 3 and 8 abuse. Once the context and interests in shifting responsibility for lethal decisions are understood, we might be less ready to regard these changes in positive law as promoting personal autonomy or advancing the interests of vulnerable patients.

The Code outlines the requirement that the patient expressly authorise consent to or refusal of life-sustaining treatment. But this does little to safeguard the patient against abuse and homicide of the kind outlined. This is because a patient will rarely know the legal ramifications of the term "treatment" nor indeed the implications of new law in permitting non-therapeutic research and, as we shall see, clinical drug trials too. There is the further requirement that certain serious healthcare and treatment decisions be brought before the Court of Protection. Those envisaged include, for example non-consensual PVS dehydration cases, organ and bone marrow "donation" cases, non-therapeutic sterilisation, abortion and research cases and other cases in which there is some dispute about whether treatment is in a particular person's best interests (paras. 6.18–6.19 Code). The very fact that cases in which there is *dispute* about whether treatment is in a person's best interests are set out as one of the kinds of case that would need to go to the Court of Protection highlights the vulnerability of patients surrounded by compliant attorneys acting on the advice of professionals.

### 4.5.3 Court Appointed Deputies

Court appointed deputies too may be involved in making healthcare decisions where "important and necessary actions cannot be carried out without the court's authority, or there is no other way of settling the matter in the best interests of the person who lacks capacity to make particular welfare decisions" (para. 8.38 Code). This will extend to "best interests" sterilisation and abortion decisions. Whether it extends to dehydration orders remains to be seen. It is explicitly recognised that deputies will often be at loggerheads with the family and that "[t]here may even be a need for an additional court order prohibiting those family members from having contact with the person" (para. 8.39 Code).

## 4.6 Other Interests: Non-Therapeutic Research, Clinical Trials, Sterilisation and Abortion on the Non-Consenting Mentally Incapacitated

Section 30 of the Act permits intrusive research to be carried out on a person who lacks capacity to consent if it is carried out—"(a) as part of a research project which is for the time being approved by the appropriate body . . ." section 31(4) (b) permits non-therapeutic research that has no potential to benefit P without P's consent provided that the research is "intended to provide knowledge of the *causes or treatment of, or of the care of persons affected by the same or a similar condition*" (the emphasis is mine).

At the same time, the *Medicines for Human Use (Clinical Trials) Regulations 2004*[14] Schedule 1 Part 5 Regulation 12 makes express provision for clinical trials on non-consenting mentally incapacitated patients upon the consent of a "legal representative". In such a case: "[i]nformed consent given by a legal representative to an incapacitated adult in a clinical trial shall represent that adult's presumed will."[15] Attorneys, advance decisions and court appointed deputies are all mechanisms by which research on the non-consenting might be achieved. The Regulations do however require that: "[t]here are grounds for expecting that administering the medicinal product to be tested in the trial will produce a benefit to the subject outweighing the risks or produce no risk at all."[16] This requirement of "benefit to the subject" is only stated in the alternative in the *Mental Capacity Act 2005*. So the 2005 Act goes much further on this point than do the *Clinical Trials Regulations*. There are, however, concerns about the way the regulations define the legal representative of an adult lacking capacity. If no satisfactory personal representative is available either the doctor responsible for the patient's care, if not involved in the clinical trial, may be the legal representative, or indeed anyone nominated by the health service body providing care for the patient. The potential for conflicts of interest and the risks to the patient presented by this possibility have been commented upon.[17]

Further, the *Medicines for Human Use (Clinical Trials) Amendment (No. 2) Regulations* 2006/2984 came into force on 12 December 2006. These Regulations amend the *Clinical Trials Regulations (2004/1031)* to allow that an incapacitated adult be included in a clinical trial if certain conditions are met notwithstanding the fact that the ordinary condition that the incapacitated adult's legal representative have given informed consent (para. 4) is absent. Regulation 2 applies where: (i) treatment is required urgently; (ii) the nature of the trial requires urgent action; (iii) it is not reasonably practicable to meet the conditions specified; and (iv) the procedure adopted has been approved by an ethics committee. The Amendment Regulations therefore allow clinical trials in emergency situations on incapacitated adults *without consent*.

If this is not enough, the background to the 2005 Act suggests a patient's very organs are at risk given the new moral and legal climate. After all, influential English-speaking philosophers have endorsed the idea of organ removal without explicit consent.[18] In 1995 there was public outcry to the draft Mental Incapacity Bill because it envisaged the removal of tissue and thus organs from the non consenting vulnerable (Clause 10 *Mental Incapacity Bill 1995*). The defence of this non-therapeutic intervention on the non-consenting mentally incapacitated may be regarded as a broadly utilitarian one. On this view, the mentally incapacitated patient, perhaps in PVS and perhaps not, is regarded a potential source of benefit to third parties and, as outlined in previous paragraphs, a "non-person"[19], one having "no meaningful life" and therefore "no best interests" morally speaking. Once the patient is regarded in this way, there can be little reason to object to use of his body for the benefit of others and indeed, in 1997 Hoffenberg et al. (pp. 1320–1321) made certain proposals in *The Lancet* in an article entitled "Should organs from patients in permanent vegetative state be used for transplantation?" The authors implied that the only reason against removing organs from PVS patients without their consent was that there was as yet no consensus in support of the activity. Accordingly, it was concluded that: "For religious, cultural and other traditional reasons, it is likely that the proposal would be rejected, nevertheless, the arguments in favour are sufficiently compelling to justify serious debate" (Hoffenberg et al., 1997, p. 1321).

But why stop at PVS patients? As a matter of fact, on the small matter of the 40% misdiagnosis of PVS outlined earlier in this paper they had this to say:

We are aware of the difficulty involved in making a correct diagnosis of PVS, and, particularly, of distinguishing the locked-in syndrome. However, in this paper we discuss the possible use of organs from those patients *in whom a decision has already been taken to withdraw treatment and allow them to die.* The actual cause of their unresponsive condition is not in this sense relevant (Hoffenberg et al., 1997, p. 1321).

They go on to claim that "if patients in PVS are thought to be sentient or capable of experiencing pain, discomfort, or distress either before or after a decision has been taken to withdraw food and fluids, a strong case could be made on humane grounds for routine administration of palliative analgesic or psychotropic therapy" (Hoffenberg et al., 1997, p. 1321). This statement simply highlights how the further end of maximising organs can often obscure a patient's very life and humanity. It also emphasises the possibility of operating with a reckless disregard for human life. It also underlines the argument against causing either death or distress to the non-consenting incapacitated. It is not an argument in favour of killing him.

I have argued elsewhere that once tests like those of "rationality, autonomy and self consciousness" are used to determine who is a "*Person" (a technical term designed by utilitarians to achieve their greater ends), the sleeping, the comatose and the drunk are indeed properly regarded "non-persons." Indeed we all go in and out of "personhood" every evening. A fuller analysis of the moral implications of personism is beyond the scope of this paper but they are signal to an understanding of the principle of equal dignity. That the non-therapeutic intervention being suggested by Hoffenberg et al., was not merely minor intervention in which the patient may have a vested interest, but serious lethal intervention in which he had no possible interest whatsoever, was nowhere discussed.

There are, to be sure, hard cases when it comes to intervention on the mentally incapacitated. A person might need a blood transfusion or bone marrow transplant and her mentally incapacitated twin might be best placed to supply this regenerable tissue. Recently, it has been recognized that an incompetent can indeed have *vested interests* in certain non-therapeutic intervention. One useful case in this area is that of *Re Y.*[20] To say that a person has vested interests in the survival of a family member is very different to the possibility of wholesale organ harvesting from the incompetent favoured by Hoffenberg et al. First and foremost the "vested interest" analysis recognizes the needs and interests of the incompetent without first stipulating that the patient has "no best interests," "no meaningful life" or is some other form of "non-person." This after all, is one of the troubling features of the personism implicit in certain judgements of the *Bland* decision and the cases that apply *Bland*. Equally, it need not be thought impermissible to undertake medical procedures (that would assist medical research in licit ways) on non-consenting adults so long as those procedures were likely to be beneficial to the patient. This would avoid the discriminatory "no best interests," "no meaningful life" personist tests advocated by the maximising theories of Hoffenberg et al. Much could be said about this possibility but for the purposes of this paper, the discussion must be limited.

To argue in favour of non-therapeutic intervention on the non-consenting mentally incapacitated as Hoffenberg et al. have done, advocating removal of vital organs in the name of social utility, highlights the commercial, scientific and other

euthanasia and a swift death by lethal injection. Accordingly, the process becomes self-fulfilling. Because "the decision (to dehydrate) has been taken," there is resultant distress to the patient and onlookers. The distress is then used as a rationale for arguing "in favour of a more expeditious mode of death, for example, administration of a lethal drug."

This was Kuhse's strategy in the mid-eighties when at the Fifth Biennial Conference of the World Federation of Right to Die Societies held in Nice, she suggested that once people accepted the removal of all treatment and care—especially the removal of food and fluids—they would upon observing what a painful way it was to die, accept the notion of the lethal injection (Marker, 1993, pp. 94, 267). Kuhse's strategy is probably well-founded. Rather than asking the question about the wisdom and justice of dehydrating "non-persons" to death, a willing populace already seduced by the language of consumerism is likely to call for state sanctioned medical killing.

Further, the decision to dehydrate a patient in turn raises Hoffenberg et al.'s question about "the possible use of organs from those patients in whom a decision has already been taken." Usefully too, the organs are fresh because the body is still alive and has not been subject to dehydration (Hoffenberg et al., 1997, p. 1321). Thus the cycle of death becomes self-perpetuating and, indeed, fuelled by interests in medical research, clinical trials, eugenics (implicit in sterilising and aborting the young of the "non-productive, unfit") and state efficiency.

To argue for the intentional dehydration of PVS patients or the administration of a lethal injection, I suggest, is to argue for the routine abandonment of the most fundamental of rights under the European Convention on Human Rights which itself came into existence precisely because these very same rights had been so flagrantly and systematically violated in 20th century Europe.

## 4.7 Other International Law

Shortly after the war, various international instruments supported a total ban on non-therapeutic research on the mentally incompetent. These included the Nuremberg Code (1947) at 1 "The voluntary consent of the human subject is absolutely essential." The World Medical Association Declaration of Helsinki: Recommendations Guiding Physicians in Biomedical Research Involving Human Subjects, adopted by the 18th World Medical Assembly, Helsinki Finland, June 1964 required that "[i]n research on man, the interests of science and society *should never* take precedence over the interests of the subject."[22] Other Covenants seeking to prohibit utilitarian invasions on the non-consenting included the International Covenant on Civil and Political Rights (G. A. Resolution 2200 (XXI), 999 U.N.T.S. 171 [1966]), article 7 which stated that "No-one shall be subjected without his free consent to medical or scientific experimentation." The World Health Organizations, Guidelines for good clinical practice for trials on pharmaceutical products (1995) WHO Technical Report series No. 850, Annex 3 at 3.3 (f) and (g) also articulate outright prohibitions on non-therapeutic research without express consent. Likewise there is hope for the vulnerable incapacitated in the Convention on the Rights of Persons with Disabilities which contains numerous re-statements and clarifications of some of the protections already mentioned: the right to life (Article 10), freedom from medical and scientific experimentation without consent (Article 15), freedom from exploitation and abuse (Article 16(5)), respect for physical and mental integrity on an equal basis with others (Article 17), retention of fertility on an equal basis with others (Article 23(1)(c)), freedom from discriminatory denial of health care or food and fluids on the basis of disability (Article 25 (f)). This Convention manifestly opens up new avenues for challenging the kinds of abuse, mutilation and homicide apparently licensed by the 2005 Act and related legislation.

It should not be assumed, however, that all international law favours the interests of the disabled. The 2000 Helsinki Declaration by contrast outlines the following:

> 2. It is the duty of the physician to promote and safeguard the *health of the people*. The physician's knowledge and conscience are dedicated to the fulfilment of this duty.
>
> 3. The Declaration of Geneva of the World Medical Association binds the physician with the words, "The health of my patient will be my first consideration," and the International Code of Medical Ethics declares that, "A physician shall act only in the patient's interest when providing medical care which might have the effect of weakening the physical and mental condition of the patient." (This latter no longer appears in the updated version of the I.C.M.E.)
>
> 4. Medical progress is based on research which ultimately must rest in part on experimentation involving human subjects.
>
> 5. In medical research on human subjects, considerations related to the well-being of the human subject *should take precedence* over the interests of science and society . . . .
>
> 26. Research on individuals from whom it is not possible to obtain consent, including proxy or advance consent, should be done only if the physical/mental condition that prevents obtaining informed consent is a necessary characteristic of the research population. The specific reasons for involving research subjects with a condition that renders them unable to give informed consent should be stated in the experimental protocol for consideration and approval of the review committee. The protocol should state that consent to remain in the research should be obtained as soon as possible from the individual or a legally authorized surrogate.

The absolute prohibitions have been removed. The duties of the doctor relate in part to "the health of the people" not that of his "patient". The International Code of Medical Ethics articulates an incoherent duty of physicians. The demands of medical progress alone are outlined in unmistakable terms.

There is now growing international support for the view that non-therapeutic research can be legitimately conducted without prior consent. This view is reflected in paragraph 4.8.14 of the 1996 guidelines of the self-styled *International Conference on Harmonisation of Technical Requirements for Registration of Pharmaceuticals for Human Use*, Article 26 of the 2000 version of the *Helsinki Declaration and the European Convention on Human Rights and Biomedicine* (the Biomedicine Convention), interpreted by reference to its Additional Protocol concerning Biomedical Research (see Council of Europe 1997, 2005). The most recent of these is the Additional Protocol to the Biomedicine Convention, which opened for signature on 25 January 2005. While the UK has neither signed the Biomedicine Convention, nor its Additional Protocols, the explanatory notes to the *Mental Capacity Act* suggest that the Act's research provisions are based on those laid down in the Biomedicine Convention (para. 96). The Convention could also have an interpretative impact on the ECHR, as given domestic effect by the *Human Rights Act 1998*. In short, Article 17(2) of the Biomedicine Convention permits research that will not benefit the participant, as long as it is intended to benefit those with the participant's condition or of the same age and satisfies certain risk requirements. The Convention thereby adopts provisions for research on all those who lack capacity that are similar to those adopted by the Clinical Trials Directive for children but not incapacitated adults. Emergency research is also addressed by the Additional Protocol concerning Biomedical Research, the provisions of which will constitute additional articles to the Convention once the Protocol is in force (Article 33 of the Protocol "Relation between this Protocol and the Convention"). Article 19 of the Protocol states that where the urgency of the situation renders it impossible to obtain consent in a sufficiently timely manner from the participant or a legal proxy, research may still take place on certain conditions. These require that research of comparable effectiveness cannot be carried out in non-emergency situations, the result is approved by the competent body, that the participant's previously expressed objections are respected, and research that is not intended to produce a benefit to the participant must seek to *benefit persons in the same population* and entail minimal risk and burden (Article 19(2)).

International law is as good as those who make and apply it. I will venture to say that it is quite possible that some of the research and intervention contemplated is indeed minor and justifiable. However, the scientistic concerns of recent conventions are unmistakable and given the utilitarian thrust of positive international law and domestic law like *Bland* (which highlights the kind of reasoning being applied in respect of patients thought "grotesquely alive" with "no best interests of any kind"), it cannot safely be supposed that the interests of science and society would not take precedence over the interests of those regarded as having none. If a patient has no best interests of any kind, then logically speaking, virtually anything may be done to him, so long, perhaps, as it does not upset onlookers. It is this logical progression of the "no best interests" argument that constitutes the mechanism by which assault, experimentation, mutilation and homicide ensue.

It is precisely because of the dubious reasoning evinced by thinkers as eminent as certain judges in *Bland*, the International Transplant Ethics Committee (i.e. Hoffenberg et al., 1997) and renowned utilitarians writing on this subject, that we cannot suppose that attempts to foster research and other financial interests would be performed in a manner consonant with the inherent dignity of all human beings irrespective of disability. Accordingly, in relation to "non-persons" or those having "no best interests" and no "meaningful life" the trumping power of illicit financial, medical and scientific interests should not be underestimated.

## 4.8 Conclusion

The experience of the twentieth century bears witness to the abuse, mutilation and homicide of the vulnerable made possible by the power of the state, mass markets, and medical and financial interests. Suggestions for reform of the law regarding food and fluids typically take place in the context of utilitarian personistic "quality-of-life" presuppositions, and interests in shifting legal responsibility for life-and-death decisions, medical research, drug trials, organ harvesting as well as more mundane bureaucratic concerns like bed-clearing. With the Western world undergoing massive demographic change and a growing ageing and non-productive population, it cannot be assumed that these alterations to the positive law are problem-free. By allowing new agents power to require that food and fluids be withdrawn, non-therapeutic research and other procedures (like abortion and sterilisation) be performed on non-consenting patients, novel legislation such as that discussed cannot be regarded as autonomy enhancing so much as a threat to human rights. These laws although touted as progressive, more often than not invite routine abuse and destruction of the vulnerable, obscure accountability and create an inconsistent body of law, with conflicting obligations for health professionals.

## Notes

[1] The *Adults with Incapacity (Scotland) Act 2004* governs the position in relation to Scotland. I concentrate on the sweeping changes introduced in England and Wales for the purposes of this paper. A broader study would consider the position throughout Britain and Europe.

[2] *Frenchay Healthcare NHS Trust v S* [1994] 1 WLR 601; *Re D (Adult: Medical Treatment)* [1998]1. FLR 411; *Re H (A patient)* [1998]2 FLR 36; *NHS Trust A v M; NHS Trust B v H* [2001] 1 All ER 801.

[3] See also: Zeman (1996, p. 144): "It is difficult to establish with certainty whether a patient is unaware, which is underlined by the high rates of misdiagnosis in PVS."

[4] Stokes, P. (1997) 'Hillsborough victim emerges from coma' *The Daily Telegraph* Thursday 27 March.

[5] In one case a young woman was able to communicate her desire to live days before a court was due to hear an application to remove her tube-feeding. Lightfoot, L. and Rogers, L (1996) 'Dead woman casts vote for right to stay alive' 7 January *The Sunday Times*; Cramb, A. (2002) 'GP who survived coma sues hospital' *The Daily Telegraph* 6 September; See also Laing, (2002, p. 1272).

[6] Sample, I (2006) 'For first time doctors communicate with patients in PVS' *The Guardian* 8 September; Owen et al. (2005 pp. 290–306) Owen et al. (2006, p. 1402) Cohen, J. (1996) 'Coma Patient Back From Dead' *The Daily Telegraph* 13 February; Anon, (1994) 'Coma Man: I was awake' *Evening Standard* 8 December; McFadyean, M. (1992) 'Lifelong Support' *The Independent* 29 November. Melanie McFadyean writes: "Mark's experiences complicate issues about the apparent quality of life sustained by people in PVS..." I could hear my friends talking, he says" I remember people saying things about me and they were wrong – I couldn't answer, of course. You feel a raging anxiety."; Toy, M. (1996) "Miracle Men" *The Sunday Telegraph* 24 March.

[7] See also *NHS Trust, A v M; NHS Trust, B v H* [2001] 2 FLR 3671, FD; *NHS Trust A v H* [2001] 2 FLR 501 (Dame Elizabeth Butler Sloss), *NHS Trust v I* [2003] EWHC 2243, Dame Elizabeth Butler-Sloss, *Re G (Adult Incompetent: Withdrawal of Treatment)* (2002) 65 B.M.L.R. 6 2001 WL

1819861 (Dame Elizabeth Butler-Sloss). In *Re G* the evidence of one expert witness was accepted both as to the severity of G's condition and as to her complete inability to recover from it. She had apparently "exhibited no meaningful response for over nine months."

[8] For a fuller discussion of personism, the idea that some human lives are worthless, a term, in any case, reminiscent of the "*lebensunwerten Lebens*" concept employed by the Nazis in mid-twentieth century Germany, (see Laing, 1996, pp. 196–225).

[9] R v *Gibbins and Proctor* (1918) 13 Cr. App. R. 134. See also Laing (1994, pp. 57–80).

[10] See the long established principle of double effect referred to in *Pretty* by Lord Bingham. He observed that the common law recognizes the principle of double effect: "Under the double effect principle medical treatment may be administered to a terminally ill person to alleviate pain although it may hasten death. . . . the case of *Bland* involved a further step . . . see also *NHS Trust v H* . . . These are at present the only inroads on the sanctity of life principle in English law." He also adds in the same paragraph "mercy killing in the form of euthanasia is murder and assisted suicide is a statutory offence punishable by fourteen years imprisonment." R (*Pretty*) v *DPP and the Home Secretary* (2002) 1 All ER 1, para. 55 per Lord Bingham.

[11] Archbishop Peter Smith and Finnis, (July, 2004). Indeed, on December 14, 2004 the intervention of Roman Catholic Archbishop Peter Smith (December, 2004) ensured that the Bill was passed in the House of Commons without amendment and without delay. Delay might have scuppered the Bill because parliament was soon to be dissolved for the coming election. *The Daily Mail* had this to say, the following morning: "It gets worse. When [Blair's] political thuggery seemed likely to backfire, he offered an apparent concession to critics. But MPs only learned of it minutes before the debate ended, when in farcical 'Parliamentary games' they were handed copies of a letter from Lord Chancellor Lord Falconer to a Catholic Archbishop setting out the terms of a possible deal. So it comes to this. Ministers refuse to compromise in Parliament, but stitch up a private understanding with a churchman, which they then use to get the Bill through unamended. It stinks. And those MPs who swallowed the party line should be ashamed of themselves. The only hope is that the Lords will give this wretched measure a mauling." In fact, the mauling in the Lords never transpired. The Bill was passed. Parliament was promptly dissolved for the election. Leader writer, (2004) 'Conscience and Abuse of Power,' *The Daily Mail*, 15 December, p. 12. See also: Ann Treneman (2004) 'No dignity in this sorry victory' *The Times*, 15 December, p. 6. Compare Laing, 2004a, p. 1165; 2004b, p. 12; 2005a, p. 11; 2005b, pp. 137–145.

[12] See also Cottingham (1996, pp. 128–143).

[13] (http://news.bbc.co.uk/1/hi/health/6353339.stm).

[14] (S.I. 2004/1031).

[15] *Medicines for Human Use (Clinical Trials) Regulations 2004* Schedule 1 Part 5 Regulation 12.

[16] Schedule 1 Part 5 Regulation 9.

[17] Nicholson, (2004 p. 1212).

[18] Harris (1992, pp. 104–107); Singer, (1994).

[19] See Singer, (1979, p. 12); cf. Laing, (1996 pp. 196–225).

[20] *Re Y* commentary (1996) *Medical Law Review* 205–207. It is useful to consider by analogy the position in relation to incompetent minors. In the US, the courts have authorised several forms of donation by minors aged seven and younger. In *Hart v Brown* 289 2 Ad 386 (1972) [29 Conn. Supp. 368, 289 A.2d 386 (Super Ct. 1972)] donation by a 7 year old to his twin was authorised. In *Cayouette v Mathieu* [1987] RJQ 2230 (Sup. Ct.) donation of bone marrow by a 5 year old to his brother was authorised.

[21] *X v Denmark* 1983 Application No 9974/82 32 DR 282.

[22] World Medical Association, *Declaration of Helsinki: Recommendations Guiding Physicians in Biomedical Research Involving Human Subjects*, adopted by the 18th World Medical Assembly, Helsinki Finland, June 1964. Recent alterations to the Declaration merely highlight the novelty of recent moves to permit what was, at one time, regarded as unthinkable.

# References

Andrews, K., Murphy, L., & Manday, R. (1996). Misdiagnosis of the vegetative state: retrospective study in the rehabilitation unit. *British Medical Journal, 313*, 13–16.

Cottingham, J. C. (1996). Medicine, virtues and consequences. In D. S. Oderberg & J. A. Laing (Eds.), *Human lives: Critical essays on consequentialist bioethics*, (pp. 128–143). London: Macmillan.

Finnis, J. M. (1993). Bland: Crossing the Rubicon. *Law Quarterly Review, 109*, 329.

Harris, J. (1992). *Wonderwoman and superman: The ethics of human biotechnology.* Oxford: Oxford University Press.

Hoffenberg, R., Lock, R. M. Tilney, N., Casabona, A. S. Daar, A. S. Guttmann, R. D., et al. (1997). Should organs from patients in permanent vegetative state be used for transplantation? *The Lancet, 350*, 1320–1321.

Kennedy, I., Sells, R. A., Daar, A. S., Guttmann, R. D., Hoffenberg, R., Lock, R., et al. (1998). The case for "presumed consent" in organ donation. *The Lancet, 351*, 1650–1652.

Keown, J. (1995). *Euthanasia examined.* Cambridge: Cambridge University Press.

Keown, J. (1997). Restoring the moral and intellectual shape to the law after *Bland. Law Quarterly Review, 113*, 481.

Laing, J. A. (1990). Assisting suicide. *Journal of Criminal Law, 54*, 106–116.

Laing, J. A. (1994). The prospects of a theory of criminal culpability: Mens rea and methodological doubt. *Oxford Journal of Legal Studies, 14*, 57–80.

Laing, J. A. (1996). Innocence and Consequentialism: Inconsistency, Equivocation and Contradiction in the Philosophy of Peter Singer. In D. S. Oderberg, & J. A. Laing (Eds.) *Human lives: Critical essays on consequentialist bioethics* (pp. 196–225). London: Macmillan.

Laing, J. A. (2002). Vegetative state – the untold story. *New Law Journal, 152*, 1272.

Laing, J. A. (2004a). Mental Capacity Bill – a threat to the vulnerable. *New Law Journal, 154*, 1165.

Laing, J. A. (2004b). Disabled need our protection. *Law Society Gazette, 101*, 12.

Laing, J. A. (2005a). The right to live: Reply to the Chief Executive of the Law Society. *Law Society Gazette, 102*, 11

Laing, J. A. (2005b). The Mental Capacity Bill 2004: Human rights concerns. *Family Law Journal, 35*, 137–143.

Marker, R. (1993). *Deadly compassion.* New York: Wm. Morrow & Co.

Nicholson, R. (2004). Another threat to research in the United Kingdom. *British Medical Journal, 328*, 1212–1213.

Oderberg, D. S. (2000). *Applied ethics: A non-consequentialist approach.* Oxford: Blackwell.

Oderberg, D. S. (2001). Starved to Death by Order of the Court. *Human Life Review, 27*, 103–112. Summer.

Owen, A. M., Coleman, M. R., Boly, M., Davis, M. H., Laureys, S., Pickard, J. D. (2006) Detecting awareness in the vegetative state. *Science, 313*, 1402.

Owen, A. M., Coleman, M. R., Menon, D. K., Johnsrude, I. S., Rodd, J. M., Davis, M. H., et al. (2005). Residual auditory function in persistent vegetative state: a combined PET and fMRI study. *Neuropsychological Rehabilitation, 15*, 290–306.

Singer, P. (1979). *Practical ethics.* Cambridge: Cambridge University Press.

Singer, P. (1994). *Rethinking life and death.* Sydney: The Text Company.

Smith, A. P. (2004). Letter to The Rt. Hon.Lord Falconer of Thoroton Secretary of State and Lord Chancellor on the subject of the Mental Capacity Bill. http://www.catholicchurch.org.uk/ citizenship/bioethics/mentalcapacity/mc041214a.htm

Smith, A. P., & Finnis, J. (2004). Briefing note on the Mental Capacity Bill (2004). July 2004. http://www.catholicchurch.org.uk/citizenship/bioethics/mentalcapacity/mc040719.htm.

Smith, J. C., & Hogan, B. (1999). *Criminal Law.* (9th ed.). London: Butterworths.

Zeman, A. Z. J. (1996). The persistent vegetative state. *The Lancet, 351*, 144.

# Part II
# Philosophers Address the Issue

# Chapter 5
# Quality of Life and Assisted Nutrition

Alfonso Gómez-Lobo[1]

In this exposition I shall first examine in broad outline the notion of quality of life and then inquire whether this notion should play a role in difficult end-of-life decisions. Finally, I shall turn to the more specific question of withholding or withdrawing Medically Assisted Nutrition and Hydration (MANH), especially from patients in Post Coma Unresponsiveness (PCU).

The expression "quality of life" can be used in different senses.[2] The words themselves evoke the idea that lives can be better or worse, i.e. that lives can be evaluated in a manner similar to the way we evaluate, for example, works of art, instruments and institutions. The idea that lives can be judged according to their quality goes back to the ancient Greeks and their efforts to discover which is the good (or the best) life. The highest quality of life they identified with happiness or human flourishing, a life lived in the enjoyment of the basic human goods.

A poor quality life, by contrast, is the life of an individual who lacks certain goods that can be of different kinds, for example, mental, physical, social, or instrumental. A mentally impaired person who suffers from a chronic disease, doesn't have relatives or friends, and lacks financial resources can be said to have a low quality of life.

Quality of life, in this sense, is a holistic notion that covers different dimensions and is therefore open to different assessments. Can there be unanimity about the quality of the life of a person who lacks certain goods (e.g. mobility) but has others (wealth)? In fact, the assessment of quality of life will vary across traditions, cultures and social groups, with some people regarding certain forms of dependency (e.g. having to be fed and cleaned) as abhorrent while others would regard them as tolerable.

One should also distinguish from quality of life a narrower concept that also admits of degrees, the concept namely of physiological condition. One's health can be better or worse. This determination of the condition of a patient is a diagnostic judgment reserved to the medical profession and is a necessary condition for effective

Alfonso Gómez-Lobo
Georgetown University
Email: gomezloa@georgetown.edu.

C. Tollefsen (ed.), *Artificial Nutrition and Hydration:*
*The New Catholic Debate*, 103–110. © Springer 2008

therapeutic intervention. Without a reliable assessment of the patient's pathological condition it is hard to see how a physician could cure her illness or alleviate her suffering, but this assessment is surely different from an over-all quality of life judgment.

Low quality of life and pathological condition are two different concepts that do not coincide and should be distinguished.

Should quality of life be a central consideration when making difficult end-of-life decisions, especially in the case of patients in PCU? The inclusion of this class of patients is of paramount importance because they can be taken to be at the lowest possible level in terms of the quality of their lives. We shall address their condition after some general remarks.

It is well known that the expectation of a low quality of life has become a standard rationalization to justify euthanasia, especially among representatives of contemporary utilitarianism. I have in mind here poor quality due to a wide variety of impediments and handicaps short of PCU. Since most utilitarians consider pain and the loss (or diminution) of the capacity to experience pleasure to be the worst evil, and since they measure quality of life by reference to pleasure and pain, it is not surprising if someone in the utilitarian camp claims that by intentionally killing a patient with low quality of life expectancy one is benefiting her. This manner of thinking is today deeply embedded in the Anglo-American culture and is spreading swiftly in the wings of globalization, first to continental Europe, notably the Netherlands, and then to other areas of the world.

In order to understand what is deeply wrong with this pattern of thought, it is useful to remember some basic notions of traditional action theory. In any action, an adequate philosophical analysis would distinguish between (a) what is done and (b) why it is done.[3] The former is the action itself (which may consist in an omission), the latter is the motive the agent has for engaging in the action, and usually consists in the expected consequences of the action. In traditional terminology these are called the *finis operis* and the *finis operantis*, the objective of the action itself and the further purpose of the agent.

Utilitarianism is a consequentialist position, and as such derives its moral judgments not simply from what an agent intends, but rather from the actual results achieved. In the case of euthanasia, the consequence is that all pain, of course, is eliminated, but we know that there are many actions that may lead to good results and yet are seriously objectionable, like obtaining peace by wholesale targeting of civilians.

An agent aiming at euthanasia can perform different kinds of action or engage in different kinds of omission that will have as a consequence the death of the patient. Can the expected low quality of life ever justify such actions and omissions? I think there are good philosophical reasons to give a firm negative reply and to assert, on the contrary, that a person in such a condition should never be denied care and respect.

In fact, although a person may be affected by a serious lack of goods, she is still enjoying the basic good of life, a good that is distinct from any evil that the person may be undergoing. Moreover, from the perspective of the person herself, life even under those conditions can be desirable although for an external observer it may appear almost unbearable. It would therefore be an intolerable presumption to judge from the outside that her life is "not worth living."

The universal prohibition of intentional killing of the innocent is grounded on respect for the dignity of a person, and human dignity is logically independent of, and not reducible to, the quality of a person's life. Dignity is an intrinsic property that does not admit of degrees. Dignity expresses the value of persons, and unlike the value of things, it does not vary. A severely handicapped person is by her very nature as valuable as anyone else, and consequently should be respected like anybody else.

There are good reasons to extend special care to people whose lives are of poor quality (they need the care more than others), but there are no good reasons to justify the obliteration of their lives because of poor quality of the latter. Quality of life considerations, paradoxically enough, do not measure the quality of life itself (for it remains an invariable basic good), but rather of other goods that may be sustained by life or be lacking from it.

To assert that life is a basic human good is not equivalent to what can be called "vitalism."[4] This position holds that human life is an absolute good that takes precedence over all other goods, and that should be preserved at all costs. The absolute value of life would thus ground stringent moral rules that should govern all end-of-life medical decisions. For vitalism the preservation and utmost prolongation of life is the sole action-guiding goal in matters pertaining to medical practice once health can no longer be restored.

In rejecting vitalism, I would affirm the traditional principle that it is rational to pursue, foster and protect goods (chiefly, though not exclusively, human life), but that it is also rational to forgo goods (in certain circumstances and in light of other goods) as long as one does not turn intentionally against them. By "forgo" I mean to give up something that one possesses or could in principle possess. An example of this attitude would be that of someone who affirms that to have children is an important human good, and yet gives up the enjoyment of this good by entering a monastery. A further case would be that of a person who, because of an insurmountable physical impediment, cannot enjoy the good of progeny. Such a person would not be giving up a good. He or she would be accepting the fact that circumstances have placed this good beyond his or her reach. This acceptance is not irrational and is perfectly compatible with a strong affirmation of the value of the corresponding good.

For vitalism, due to its faith in technology and its frontal resistance to the forgoing of life, the idea of the acceptance of death is a source of anger and rebellion. This is an attitude that is sometimes found, I am told, among relatives of cancer patients who want the physician to continue chemotherapy after it has ceased to be effective and long after it has begun to show its devastating effects on the dying patient. Illusory hopes or the arrogant determination to defeat illness and death replace the deeply human attitude of acceptance of the inevitable ending of our lives.

For those who stand on the middle ground between vitalism and thanatism, between the will to preserve life at all costs and the will to cause death when deemed convenient, an intractable dilemma arises in certain difficult cases, most notably in the case of patients in PCU.

First a few points about the terminology used. The standard way to refer to this condition is "Persistent Vegetative State" (PVS) or more generally in a "Vegetative State" (VS). These labels are remotely based on Aristotle's distinction of faculties or powers (*dunameis*) in living things (*De Anima* II. 3. 414a 29–415a 13). In these patients the rational and sensitive powers of the soul can no longer be activated in the normal way and thus what is left exercising its activity is only the *threptikón*, the "vegetative" power, that is independent of consciousness and the will. However, the term "vegetative" easily leads people to think of these patients as "vegetables", and thus as non-human. In order to avoid this terrible misunderstanding, a new label has been proposed "Post Coma Unresponsiveness" (PCU) which is free from the misleading associations, and I am happy to adopt it.[5]

The Aristotelian framework, on the other hand, provides apt conceptual tools to reject the view that patients in PCU have ceased to be persons.[6] Aristotle (*De Anima* I. 4. 408b 18–25) expresses the common sense view that each of us is a single organism, a single, fully unified substance, and that the loss of the higher faculties is not the passing away of a substance, called "the person", residing in a different substance called "the body", as contemporary dualists hold.[7] The loss of consciousness and reason is the loss on the part of a person of her capacity to activate her higher powers due to physiological impairments, but is not equivalent to the death of the person. Although a patient in PCU may show no sign of responsiveness to his environment, this is not a reason to deny that she is the same person that was conscious before she suffered the trauma that led to PCU. It follows that she now deserves the same respect she deserved at earlier stages of her life.

The difficult dilemma I mentioned earlier arises out of the fact that most PCU patients cannot swallow because of their pathological condition and hence require assisted nutrition and hydration (ANH), also referred to as medically assisted nutrition and hydration (MANH). What can moral philosophy say about the proper way to decide whether to provide, withhold or withdraw such assistance?

There are many factors that should help us realize that it is very difficult to provide a univocal answer that would apply to all cases. There is in the first place the difficulty involved in deciding whether nutrition and hydration by means of a nasogastric tube or a gastrostomy tube constitutes normal care or medical treatment. Once a tube has been inserted, the feeding can be done by an untrained person, but the decision to insert a tube, the insertion itself (especially when surgery is required), the prescription of the food and liquids to be provided, the monitoring of the patient and of any side-effects (aspiration, infections, etc.) clearly require the skills of a physician.[8] Since the day-to-day feeding, which is the ordinary care part of the process, is only possible because of the prior, medically accomplished insertion of the feeding tube, it seems to me, as matter of logic, that the procedure as a whole should be deemed to be "medical treatment." In fact, a medical intervention provides the *necessary* conditions for the inception of the ordinary care, and medical monitoring is part of a responsible continuation of the care. In a village without doctors, ANH of patients in PCU is impossible.

The classification of ANH as medical treatment does not, by itself, resolve the ultimate moral question whether its provision is obligatory or optional. It does help to sharpen the issue by inviting us to consider the traditional criteria for inception, continuation or cessation of treatment, namely its benefits and burdens.[9]

In order to discuss the application of these criteria to MANH I would like to consider an extreme hypothetical claim, the claim that withholding or withdrawing MANH is always and necessarily euthanasia by omission. In order for this claim to be true there would have to be a logical or conceptual connection between the *finis operis* and the *finis operantis*, in other words, *the action would have to be such that it could only be performed by an agent with a specific further intention*. There would be no room for exceptions.

I submit that such a connection exists between an act of active killing and the intended goal of euthanasia. A physician who gives a patient a lethal injection surely intends the death of the patient. The act of itself would not be performed were it not for the intention of the agent. That further intention is a necessary condition for the performance of an action that directly causes the intended effect.

Omissions are different. They do not actively cause death. They usually remove an impediment for the operation of a different cause. When a respirator is turned off and a patient dies, what causes the death is the prior acute pathological condition, not the actual unplugging of the machine. A patient in a less serious condition who has been put temporarily in a respirator would go on living. Turning off a life-sustaining machine, by itself, does not cause death.

Due to the manner in which causality operates, an omission may or may not be linked to a specific intention. *If the connection between the omission and the intention were not contingent but necessary, then every withdrawal of treatment or of artificial life support would have to be considered euthanasia.* In traditional language this would entail that all means should be considered ordinary means. If a doctor withholds antibiotics from a patient who has been in PCU for several years, the doctor could be suspected of intending the death of the patient. If a doctor halts chemotherapy at the request of a patient, he could be accused of engaging in physician-assisted suicide.

These clearly unacceptable accusations are meant to show that the connection between the *finis operis* and the *finis operantis* in the case of omissions is indeed contingent. That there are legitimate forms of letting die is central to the position that values life and yet accepts the forgoing of life, that strives to care for the sick, but rejects excessive medical treatment.

It is true that certain omissions can be the result of intending the death of the patient, but they need not be. For an external observer there may be no perceptible difference between euthanasia by omission and legitimate letting die because the intention with which similar actions are performed cannot be judged from the outside. If a dedicated physician examines carefully whether a treatment is futile or burdensome, or both, and decides in good conscience to halt it with the intention of removing the burden and hence letting the patient die of her underlying illness, it would be presumptuous to impute to him the intention of causing the death of the patient and thus of engaging in euthanasia.

The same holds, I conjecture, for certain cases of PCU.[10] There are instances in which an underlying pathological condition prevents the patient from taking food normally so that it is precisely the pathology that calls for the surgical insertion of a gastrostomy tube. Again it would be presumptuous to blame a physician if, after a lengthy period of time and after considering the burdens on the patient herself and her family, on the surrounding community, as well as a host of other factors (such as lack of health insurance), he recommends that the medically assisted nutrition and hydration be forgone.[11] We cannot know what someone's ultimate subjective intentions are, but we should not assume that the doctor is necessarily intending the death of the patient. As I have argued, from the point of view of action theory, there is no such necessity. The recommended action may be a case of letting die for legitimate reasons because he has concluded that it would be disproportionate to continue. The primary intention of the action in that case is to relieve a burden. This does not presuppose that the patient is considered somehow "worthless" or "unproductive" or lacking in dignity. Consideration of the burdens on the patient, the family and the community are legitimate reasons to decide.

Although I have focused on the burden component of the arguments for letting die, a word is due on the futility component. It is generally accepted that if a treatment is futile, that is, if it does not achieve the appropriate goal (for example, when chemotherapy no longer works), it is morally permissible to stop it. This does not entail that it is obligatory to stop it. There may be cases in which, say, for psychological effect on the patient or the family, it may be legitimate to carry on, but only if the treatment is not significantly burdensome to the parties involved.

Until now the proper way to consider futility has been to restrict it narrowly to the means to be decided on. In the case of MANH the question is whether or not providing nutrition through a gastrostomy tube is achieving the goal of nourishing the patient in spite of his or her over-all pathological condition. But in the future we will increasingly face a new challenge. By its very nature, modern medicine strives to develop more and more efficient methods (i.e. less and less futile methods) in all fields, and also in MANH. The net result will be that in spite of their serious pathology and a hopeless prognosis certain patients will be kept alive for longer and longer periods of time.[12] Many thoughtful people consider it unreasonable to prolong life just because we have the technology that allows us to do it. The new technology, in the restricted sense, is not futile. It is certainly efficient, but in a more holistic approach to futility one could say that the technological advances are forcing us to act in ways that are futile from a broader perspective. Isn't there an element of futility in keeping a patient alive for 10 or 15 years if there is reasonable assurance that she will not recover from a condition that keeps her unconscious and unable to swallow?

I do not think we are able to articulate yet this broader concept of futility within the traditional framework, but I suspect we will be led to do so precisely because modern medicine with its quest for efficiency (and its eagerness to diminish burdens and side-effects) *will force us to consider virtually all means as ordinary*. The paradoxical and unpalatable consequence of this is that it will have become morally impermissible *not* to use the latest technology, precisely because it will increasingly satisfy the restricted application of the criteria: strictly speaking the means themselves will be neither futile nor burdensome. However, the pressing need to rethink our traditional concepts lies beyond the scope of this paper.

In summary, I have argued here for a minimal claim based on the traditional criteria for optional and obligatory means. I have acknowledged the advance made by "the culture of death" in its efforts to legitimize euthanasia, and also the dangers derived from the progress of medical technology that allows for the virtually indefinite prolongation of life. I have tried to steer clear of both thanatism and vitalism. My conviction is that care for the handicapped and the unconscious should always be our primary concern, an obligation that requires no justification because it is deeply rooted in our very humanity. The suffering and the weak have a special claim on us. The quality of the life of the patient should not affect this primary obligation. Life continues to be a basic good to be protected. But I have also argued that we should not assign morally unacceptable intentions in those cases in which conscientious physicians and families, after careful deliberation in light of the traditional criteria, opt for the withholding or withdrawal of nutrition and hydration of a PCU patient. Their action need not be interpreted as removing nutrition in order to end the life of the patient in order to end the burdens. It seems to me possible to end the MANH without intending the death of the patient. The primary intention would be the relief of an evil for the family and the unintended, yet foreseeable effect would be the death of a severely afflicted patient kept alive by means that have come to be regarded as extraordinary in the circumstances.

In sum, I have tried to show that legitimate instances of letting die in some cases of withdrawal or withholding of MANH are possible.

## Notes

[1] The author is a member of the U.S. President's Council on Bioethics. The ideas presented here are his personal views and do not represent the positions of the Council. An earlier version of this paper was delivered at the Eleventh Assembly of the Pontifical Academy for Life, Vatican City (Feb. 2005). I am grateful to Christopher Tollefsen for criticism that led to various changes in the original draft.

[2] For a collection of documents and studies see Walter and Shannon (1990).

[3] This analysis is further developed in Gómez-Lobo (2002, pp. 48–56).

[4] This use of the label is relatively new and should not be confused with vitalism as a doctrine in the philosophy of biology upheld by Bergson, Driesch and others. Cf. Beckner (1967). On the importance of considering vitalism I owe much to Keown (2002).

[5] Cf. Boyle et al. (2004). PCU could be confused with Minimal Consciousness State (MCS) and hence its correct diagnosis is of decisive importance.

[6] For a succinct expression of this widely shared view see Brody (2003, p. 281) "... life support could in these cases be unilaterally withdrawn when the organism no longer composes a person because the cortex no longer functions."

[7] It should be kept in mind, however, that Aristotle does not think in terms of personhood, a concept that was developed much later.

[8] Cf. American Academy of Neurology (1990, especially section II.C.1).

[9] Pope Pius XII (1999, pp. 213–214).

[10] In this short (and insufficient) treatment I am assuming that the will of the unconscious patient is not known. For Pope Pius XII the presumed will of an adult patient was the foundation of the rights and duties of the doctor and of the family. Cf. Pope Pius XII, *ibid.*

[11] Apart from the direct physical risks to the patient, such as risk of perforation of the bowel, infection around the opening, aspiration pneumonia, etc. the emotional strain on the caregiver, especially if he or she is a close family member should not be minimized. The burden can be increased considerably if the family resources are limited and the same persons have to carry the whole weight of paying for the hospitalization, nursing home services or home care in countries where there are no adequate public health services.

[12] The longest recorded instance of survival in coma in the US is that of Elaine Esposito who lived in an unconscious state for more than 37 years. Cf. Walter and Shannon (1990, p. 86 n. 8.)

# References

American Academy of Neurology. (1990). Position of the AAN on certain aspects of the care and management of the persistent vegetative state. *Neurology* 1989, *39*, 125–126 In J. J. Walter & T. A. Shannon (Eds.), *Quality of Life. The New Medical Dilemma.* New York: Paulist Press, 191–194.

Beckner, M. O. (1967). Vitalism. In P. Edwards (Ed.), *Encyclopedia of Philosophy*, 8. New York: Free Press, 253–256.

Boyle, J., et al. (2004) *Reflections on Artificial Nutrition and Hydration* [On line]. Available at: www. utoronto.ca/stmikes/bioethics.

Brody, B. (2003). How much of the brain must be dead? In B. Steinbock, J. D. Arras, & A. J. London (Eds.), *Ethical Issues in Modern Medicine.* New York: McGraw Hill, 277–282.

Gómez-Lobo, A. (2002). *Morality and the Human Goods. An Introduction to Natural Law Ethics.* Washington D.C.: Georgetown University Press.

Keown, J. (2002). *Euthanasia. Ethics and Public Policy. An Argument against Legislation.* Cambridge: Cambridge University Press.

Pope Pius XII (1999). The prolongation of life (Nov. 24, 1959). In K. O'Rourke (Ed.), *Medical Ethics: Sources of Catholic Teachings* (3rd ed.). Washington DC: Georgetown University Press, 213–214.

Walter J. J. & Shannon T. A. (Eds.). (1990). *Quality of Life. The New Medical Dilemma.* New York: Paulist Press.

# Chapter 6
# Towards Ethical Guidelines for the Use of Artificial Nutrition and Hydration

Joseph Boyle

ANH is an acronym for "Artificial or artificially provided nutrition and hydration." I take it that this expression refers to efforts of care givers to provide nourishment and hydration to a person independently of that person's own activities of eating and drinking. ANH is used when a person cannot eat or drink or when a person's eating and drinking is in some way problematic, for example, when efforts to feed the person are expensive or unsuccessful.

The widespread use of ANH for people in various conditions has raised ethical questions concerning the conditions under which ANH should be used and not used, including conditions which might warrant discontinuing the use of ANH, as conditions change after it was appropriately initiated. These questions indicate the need for guidelines to assist patients, families, health care and other helping professionals in addressing the moral complexities of this increasingly common choice.

Different moral communities and traditions will approach the task of formulating such guidelines rather differently, developing the resources of their traditions in various ways. The guidelines towards which this essay points are those of the Roman Catholic tradition.

The Catholic tradition has much guidance to offer concerning health care decisions, especially those that may shorten a person's life or seriously affect his or her vocational commitments. That guidance is found in the principles and casuistry of Catholic moral teaching, and is sometimes highlighted by specific moral teaching by popes and bishops (John Paul II, 1995; *Catechism of the Catholic Church*, 2268–2283, 2288–2296, 2299–2301, 2318–2325; National Conference of Catholic Bishops, 1994). In the case of the use of ANH, there is not only the traditional teaching about ordinary and extraordinary treatments and the complex teaching about euthanasia (John Paul II, 1995, paragraph 65, 712), but also a very recent papal teaching on the use of ANH for patients in what is somewhat unfortunately called "persistent vegetative state" (John Paul II, 2004). Although the condition of

Joseph Boyle
University of Toronto
Email: jboyle@chass.utoronto.ca

C. Tollefsen (ed.), *Artificial Nutrition and Hydration: The New Catholic Debate*, 111–121. © Springer 2008

persistent vegetative state occurs rarely, and although the features which complicate decisions about the use of ANH for patients in that condition are not common to most decisions about ANH, a careful look at the papal statement is a necessary step in the undertaking of formulating Catholic guidelines and a useful enter into that effort.

## 6.1 Pope John Paul II's Address on the Persistent Vegetative State

### 6.1.1 A Brief, Analytical Summary

This address is a short speech made up of seven numbered paragraphs.

In the first paragraph, John Paul greets the group of experts gathered by the Pontifical Council Pro Vita to discuss the ethical issues that arise in caring for those in persistent vegetative state. He announces the focus of his talk: the "the complex scientific, ethical, social and pastoral implications" of the condition of patients in PVS. So, the focus is PVS and not ANH except as used for those in this condition (John Paul II, 2004, 739). In the second paragraph, John Paul makes explicit what he takes the clinical and neurological reality of PVS to be.

His understanding of PVS is based on the relevant clinical and scientific evidence: John Paul defines the vegetative state as the absence of evident signs of self-awareness or of the environment, and the apparent inability to interact with others or to specific stimuli; he also underlines a morally important implication of the condition: those in it are completely dependent on others.

Before considering what makes the vegetative state so defined "permanent" or "persistent," John Paul reflects on the diagnosis and prognosis of the condition. Diagnosis is quite difficult, as the high number of reported diagnostic errors indicates. The prognosis for those given appropriate treatment and rehabilitation is mixed: some have emerged from the condition, but others remain in it "even for long stretches of time and without needing technological support." Later in this paragraph he suggests we lack the resources to distinguish patients having these very different prospects, even among those whose vegetative condition persists.

He next considers the persistence or permanence of the vegetative condition of some patients. The persistence of the vegetative state of a given patient is simply the fact that it continues for a year. The significance of this duration is not the emergence by a year of a medical factor that would change the diagnosis, but rather the fact that, as a matter of statistical summary, patients who remain in vegetative state for a year are very unlikely to recover from it, and that that low probability gets progressively lower as time goes on.

John Paul goes on immediately to note that even some patients who remain in vegetative state for many years recover from the condition. He concludes that medical science cannot predict with certainty who, among patients in this condition, will recover and who will not (John Paul II, 2004, 739).

In paragraph 3, John Paul states in general terms the Church's normative position on how people in this and other similar conditions of grave debility are to be treated. He begins this paragraph by reflecting on the reference of the term "vegetative" in the widely used, but unfortunate name for this condition. It refers not to the patient, who remains a human being, but to the clinical state. If taken as referring to the sick person as such, this expression actually demeans the person's value and personal dignity. This central affirmation is that a human being remains that as long as she or he is alive, however debilitated his or her condition: "I feel the duty to reaffirm strongly the intrinsic value and personal dignity do not change, no matter what the concrete circumstances of his or her life." This means that a person who is disabled in the exercise of higher human functions does not become an animal or a vegetable. So, human dignity remains for these people: "the loving gaze of God the Father continues to fall upon them, acknowledging them as his sons and daughters, especially in need of help"(John Paul II, 2004, 739).

In the fourth paragraph, John Paul applies the general considerations in paragraph 3 to the clinical reality described in paragraph 2. He first addresses the general rights of persons in PVS and then considers the specific issue of the use of ANH for these patients. The rights of these patients are correlative with duties of health care professionals, society and the Church, the non-fulfillment of which will lead to a failure of solidarity and of professional ethics.

He lists the rights of those in PVS, "awaiting recovery or a natural end": the right to basic health care, including such things as nutrition, hydration, warmth and cleanliness: the right to the prevention of complications related to confinement to bed; and the right to appropriate rehabilitative care and monitoring.

Turning to the use of ANH for these patients, and in particular to the decision to withdraw ANH, John Paul begins with two distinct normative affirmations: first, the use of ANH for these patients is a natural means of preserving life and not a medical act. No explicit conclusion is drawn immediately, but the context suggests this is given as a reason for holding the continued use of ANH to be obligatory.

The second point is logically independent from the first, connected to it by the conjunctive "furthermore." It is that the use of ANH in such patients is "in principle ordinary and proportionate and as such morally obligatory, insofar as and until it has achieved its proper finality, which in the present case consists in providing nourishment to the patient and alleviation of his suffering."

He goes on in a statement that appears to draw together into a conclusion the two preceding points: normal care due to the sick in such cases includes the provision of hydration and nutrition. He then observes that this obligation of minimal care is not overridden by the judgment that recovery from PVS is very unlikely.

John Paul then develops what appears to be an independent line of argumentation against withdrawing ANH from these patients: that if done on the basis of the dim prospects of recovery, it is a case of euthanasia by omission. He begins by noting that the inevitable outcome of withholding ANH from these patients is death by starvation or dehydration. He then notes that if done knowingly and willingly, the decision to withhold ANH, presumably on the basis of this judgment about unlikely recovery, is euthanasia by omission.

He wraps up this central paragraph with a reminder that doubts about the presence of a human life are to be settled in favor of the life, and so the possible presence of human life precludes actions aiming at anticipating its death (John Paul II, 2004, 739–740).

The fifth paragraph addresses the challenge posed by "quality of life" considerations, which John Paul thinks are often dictated by outside psychological, social and economic pressures. He notes that such considerations cannot override the moral principles articulated and applied in paragraphs 3 and 4. He spells out three specific implications of this general ethical observation: first, costs cannot outweigh the good of human life. Second, no external assessment of the quality of a person's life can be a ground for compromising human dignity. Any such ground introduces a discriminatory and eugenic principle into social relations. Third, the withdrawal of ANH might well cause considerable suffering on the part of those in PVS (John Paul II, 2004, 740).

In the sixth paragraph John Paul affirms the social obligation to take positive steps to oppose the withdrawal of nutrition and hydration, especially support for the families of people in this condition. He takes note of the "heavy human, psychological and financial burden" borne by the families of those in PVS, and concludes that support is necessary. Although the care of these patients is not, in general, particularly costly, society should make resources available for the care of this kind of frailty (John Paul II, 2004, 749).

In the final paragraph John Paul highlights his conclusion: the medical profession is "to cure if possible, always to care." And he provides its ground in the Lord's words: "Amen I say to you whatever you did to these least brothers of mine you did for me" (John Paul II, 2004, 740).

### 6.1.2 Some Comments

As his parting comments highlight, John Paul thinks that the provision of ANH to those in PVS is a requirement of normal care, and is so far forth obligatory. He recognizes that this obligation can be onerous, at least for the families involved. He does not say, but natural law implies that the particular forms of the duty to care are inevitably limited by the responsible agent's capacity to provide care, not least by the other grave responsibilities the person may have. For some poor families without the third party support of socialized medicine or insurance, there might be very little they could afford by way of medical care for a debilitated family member. But that does not remove their duty to care; it changes its shape.

Likewise, the obligations of political society to guarantee support for people suffering various forms of grave debility and for their families are limited by other social responsibilities. Of course, the rationing of health care and of other resources that is, in the world as it actually exists, a virtual necessity will be morally proper only if it is done in accord with the principles of fairness. Those principles exclude putting a price on a human life, but they do not exclude using scarce resources for some and not for all who have need of them. And the Pope seems to be arguing that we will not treat fairly those in PVS unless we continue to care for these patients, and that that includes maintaining the ANH that was in place as they developed this condition. So, it is morally significant that the costs of care for these patients are not high. For if they were, the requirements of fairness might be different and the form taken by the social duty to care might vary in ways similar to those in which an unsupported family's duty varies.

But plainly John Paul thinks that there are other morally compelling considerations at play in these cases besides the requirements of social solidarity, namely, that providing ANH to these patients is in principle ordinary means and so obligatory, and that withdrawing it is at least sometimes euthanasia by omission.

I understand ordinary treatments as those not involving grave burdens.[1] Burdens are the bad side effects of treatment, understood broadly. If these are either insignificant or not sufficiently important to justify making their avoidance the basis for refusing the treatment they are not grave. Extraordinary treatments, by contrast, are those having negative side effects so weighty that a person can reasonably choose to avoid those side effects, and so reasonably can decline the medical treatment. If a person chooses to reject ordinary treatments, that implies either that he or she is deliberately rejecting the good promised by the treatment, for example, continued life, or is insufficiently sensitive to the requirements of the good of life and health, perhaps from irrational fear of treatment or simple distaste for its rigors.

On this understanding of ordinary and extraordinary treatments, the nature of the standards for determining the bad side effects of the treatment to be gravely burdensome requires some articulation. I believe that the standard is ultimately the configuration of values and priorities set within a person's unique vocation (Boyle, 2002). That standard allows for the wide diversity of judgments about what is gravely burdensome that Catholic thought and practice accepts, without any suggestion that it is simply a matter of personal preference.

On this fast account of ordinary and extraordinary treatments, John Paul's judgment that providing ANH to those in PVS is in principle ordinary treatment is intelligible. To consider such a question "in principle" seems to me to mean that the only factor about a patient's situation to be considered is the fact that he or she is in PVS, and the only aspect of care to be considered is the provision of ANH. To judge the result of that consideration ordinary treatment is to judge that the negative side effects on a person in PVS of the action of providing ANH are not so serious as to justify withdrawing that treatment. Thus understood, John Paul's judgment is clearly correct.

One reason for this is that this is that providing ANH by a gastric tube seems to have relatively few negative side effects, especially for those in PVS, and even these, we suppose, cannot be experienced by the patient. Moreover, as the Pope notes, the costs for others, broadly construed, also seem generally low. And if there is a significant cost it is the cost of the care as a whole and not of the ANH alone (Boyle, 1995).

If John Paul means something like this, his teaching does not exclude the possibility that there might be burdens sufficiently serious to justify abandoning the level of care that involves those burdens; but abandoning a certain level of care–for example, in a hospital or nursing home because of a family's inability to pay–does not mean abandoning all care, and the need to move to a lesser level of care does not by itself justify withdrawing ANH. For one reason to withdraw ANH as part of a decision to lower costs might be to lower them by choosing to end the patient's life: costs go down because the patient dies, and stopping ANH contributes to the cost cutting because of its result in death. That is similar to the concern about euthanasia that John Paul discusses.

Before turning to that issue, however, the condition John Paul places on the obligation to continue ANH in these patients should be noted. He says this obligation holds "insofar as and until it is seen to have attained its proper finality, which in the present case consists in providing nourishment to the patient and alleviation of his suffering." Plainly this qualifier limits the obligation to cases where nourishment and alleviation of suffering are physically possible, and equally plainly it excludes as morally relevant changes in the diagnostic situation of these patients which do not remove the availability of these ends. I do not think that John Paul is here overlooking the dynamism of the clinical and diagnostic situation, or rejecting the possibility of withdrawing a form of care such as ANH. He is suggesting, however, that unless that dynamism reveals new or greater burdens that render the treatment gravely burdensome in changed circumstances, there will be continuing duty to treat.

What precise connection does John Paul make between the decision to withdraw ANH from a patient in PVS and euthanasia by omission? I take it that the idea of euthanasia by omission does not refer to all decisions to withdraw a treatment believed necessary for continued life (and so thought by the decision makers sufficient or likely sufficient, given other causal factors already in play, to bring about death). For, plainly, many choices to forego extraordinary treatments are of this kind. The choice to withdraw artificial respiration is often related to the patient's dying in exactly this way. Moreover, some actions that shorten life or inevitably lead to death are justifiable according to double effect (Boyle, 1980).

So, an intentional condition is required.[2] The moral meaning of the choice not to do something or to stop doing something is specified by what goal that serves and by precisely how the choice to stop serves it. In euthanasia, the aim is to stop the patient's suffering by ending the patient's life. So, euthanasia by omission is a choice to stop a life sustaining or extending treatment so that the person dies and suffers no more. Thus, if the answer to the question: "what does stopping treatment have to do with the goal one seeks?" is "Stopping the treatment will lead to the person's dying (or dying sooner) and that will stop the suffering or end the indignity etc" then there is euthanasia (or at least intentional homicide) by omission.

If something like this is the idea of euthanasia by omission which John Paul is employing, then one can see how withdrawing ANH from those in PVS can be euthanasia by omission. I suggested a version of this above, where death is chosen as a means to end expensive and burdensome care, and the withdrawal of ANH is the means to ending life. But John Paul may be read as saying that any choice to withdraw ANH from a person in PVS is euthanasia. I suggest that this is a misreading.

As I already suggested, euthanasia by omission is not any withdrawal of treatment that leads inevitably to death but withdrawal or refusal of treatment involving the intention to end life, for the sake of ending life. The rationale that John Paul has in his sights here is that founded on the assessment of the dismal future prospects for recovery of a person in PVS. Implementing that rationale by a decision to withdraw ANH is difficult to formulate without including the patient's death as part of the plan. That is clear if we ask what is it that practically connects concerns about these dim future prospects and a choice to end ANH? The patient is unlikely to recover, but continues to live. That condition is taken as bad, as indeed it is. One way to avoid that bad condition's obtaining is to bring the person's life to an end; withdrawing ANH virtually guarantees that, and this life ending aspect of the choice to withdraw ANH provides the only connection between it and dealing with the repugnant prospect of living on in PVS with ever more limited prospects of recovery. Consequently, if that line of thinking is implemented, the person's death is intended and the choice to stop ANH is, if not euthanasia strictly speaking, then intentional homicide by omission. But plainly, this is not the only possible rationale for discontinuing the ANH. Nevertheless, although I do not think that the Pope should be understood as holding that the withdrawing of ANH from those in PVS is as such euthanasia by omission, two quite common motivations for withdrawing ANH from those in PVS do involve a homicidal intent: namely, getting the treatment or the care over with, and avoiding the repugnant condition of life without recovery.

But it may seem that living the life of such a patient is more than simply repugnant; it is somehow intolerable. Preserving such a life and caring for a person in that condition can seem worse than meaningless; it appears to be an absurd suffering imposed on caregivers by a moral logic that has become divorced from reality.

The Pope plainly recognizes the real tragedy of those in PVS and the real burdens of those who care for them. But I think he would resist the idea that caring for those in PVS is absurd suffering. Our way of participating in God's loving providence is by pursuing the human goods according to the light of the principles of the natural law. That is what we have to guide our lives, not the full story of God's providence, but just what we must have to participate in God's plan as rational creatures. If we make good choices we have done our part, and the suffering that ensues because those choices are sometimes of actions that are deeply unsatisfying and troubling is not absurd but justified, though we can only trust that the justification is there.

Sometimes we get a hunch or are given a glimpse of that justification, but for many of us, much of the time, the suffering that trying to do what is right leads to is a terrible slog that faith alone can sustain. Keeping people alive who likely will never again experience, think or interact with others seems absurd to many people, but it is not absurd if we think we cannot stop this effort without treating them unfairly, or without breaking faith with them, or even without intentionally ending their lives. The goods realized by refusing these choices are real, and even in our suffering we can experience them. And if the only way we can see through such situations is to do what continues our suffering, we must accept it.

## 6.2 What Guidance does the Papal Statement on PVS Provide for Other Decisions About ANH?

I have emphasized in my exposition of and comments on John Paul's address that his focus is primarily on the care of those in PVS, and not on ANH as such. I also emphasized that much of the moral vocabulary used by John Paul is context and circumstance sensitive. What is part of the normal care all are entitled to receive must vary as the capacities of providers and their social supports vary. What is ordinary treatment often does not emerge in the abstract description of a condition and its treatment but appears only in circumstances that do not distinguish acts into generally described kinds. And the rationales in which the intentions of actions and omissions are revealed can vary considerably even though they rationalize clinically identical decisions.

All this suggests at the very least that the papal statement contains principles and patterns of moral reasoning that are widely applicable to decisions about ANH. And there may be a more direct application of the papal teaching to decisions about the use of ANH in patients not in PVS. Certainly, in cases that are quite similar to PVS in their morally relevant features, the Pope's conclusions directly apply. In particular, in cases where the patient is not dying as long as ANH and comfort care is provided, but where the patient is not capable of conscious interaction with the environment, and where recovery from that condition is most unlikely, then the papal teaching about PVS fully applies. For example, if those with Pick's disease are for a time in a PVS like condition, then removing ANH would be excluded, taking into account the traditional provisos I mentioned in the preceding section. But many of the choices about the use of ANH concern patients who are not unconscious, or whose prospects of recovery are real, or who may be able to swallow on their own, or who are so far along in the dying process that the benefit of extending life by ANH is so slight that if assessed with a realistic acceptance of impending death it would not be used.

For example, the use of ANH to assist people in various conditions of dementia seems quite different from its use for those in PVS. Perhaps these people can eat and drink. If so, deciding not to start or to stop ANH for which there are medical indications, is not potentially euthanasia by omission. Moreover, there appear to be medical indications contrary to those calling for the use of ANH, for example, evidence of increased morbidity and mortality caused by the use of gastric tubes. So here, even if ANH is considered part of normal care, it must be assessed in terms of the burdens and benefits analysis of the ordinary/extraordinary means doctrine.

Moreover, many of these people, and many others for whom ANH may be indicated, can experience the effects of the activities involved in ANH, and may well manifest a level of distress in going through this process that warrants the judgment of others that they are suffering a grave burden.

My conclusion is that the specific papal teaching about withdrawing ANH from patients in PVS is not readily exportable, and that it certainly does not exclude all choices to withhold or withdraw ANH. What are exportable, and should underlie any Catholic guidelines on ANH are the principles that underlie the papal teaching. The chief of these has application far beyond the care for people in PVS, namely, the principle of human dignity that requires respect for the human person and insists that all human beings, however debilitated, are human persons. This general principle has three more specific implications that are widely applicable to decisions concerning ANH: (1) a principle of solidarity or care for others, especially those most in need, that challenges us to reach out to those in need and to be fairminded in our attitudes and use of resources for the benefit of the most disabled: (2) the doctrine of ordinary and extraordinary treatments that reminds us that the duty to act for our own and others' life and health is serious but defeasible in the face of other grave obligations, especially those of each person's unique personal vocation; and (3) the absolute prohibition of deliberately choosing to end human life, by act or omission, for any reason whatsoever, however beneficent that reason may be. This norm is not self applying, especially in the case of choices to refrain from or stop doing something that has some human point, since behaviorally similar choices not to do something can be very different moral acts depending upon one's reason for choosing.

In any human community shaped by what John Paul has called the culture of life, these three implications of human dignity will play an irreducible role. I suggest that they are a necessary ingredient in any set of guidelines for such a community. But none is a simple rule; each requires moral thinking, sometimes difficult moral analysis. I don't think that a set of guidelines made of easy-to-understand rules can capture this moral reality.

## 6.3 The Role of Advance Directives in Decisions Concerning ANH?

The right of competent patients to refuse health care is widely accepted in modern societies. The extension of that right into the future to settle treatment issues that might arise after that person becomes unable to make decisions has also become widely accepted, in the form of living wills and the appointment of proxies with the power to make decisions on one's behalf. Since this right and its extension have become part of the social fabric of modern health care, guidelines concerning decisions to use, decline or stop using ANH must address the role of advance directives.

To begin I will give a very brief rationale for the right to refuse medical treatment and for its extension in advance directives. This rationale is not the popular secular argument which understands autonomy as an overriding value in health care, and interprets it as implying that patients' desires establish what is right and wrong. Rather, it is an argument to the effect that the right of a competent person to refuse medical treatment is a generally reasonable allocation of decision-making authority in the social realm of heath care. The patient must live with the regime of treatment and its consequences; they affect his or her vocation and life prospects most immediately. Health care professionals are not so directly affected and they are not public authorities but providers of expert advice and assistance. Such considerations as these seem to me to underlie older views about the right of competent patients to refuse treatment, the sort of views held by Pius XII, for example.[3]

The idea of social authority is that when there is disagreement about common action, as health care must be, someone must have the final say, and that person in the health care setting is most reasonably the patient him or herself. This allocation of decision making authority does not imply that the patient's views are sound or altogether morally upright–only that they will prevail if there is disagreement. This is hardly patient autonomy in the modern sense.

If this right is accepted then its extension by devices such as designating some-one with powers of attorney in respect to one's health care is also reasonable, since one may, when unable to decide for oneself, face health care prospects in which one might be required to suffer unreasonably, or to impose on others in ways one can reasonably reject, or to use one's own or other's resources in ways one would prefer to avoid.

If some such rationale for the right to refuse treatment is correct, then advance directives are a reasonable device. The question that arises in respect to ANH is whether any antecedent refusal of continued ANH must necessarily be a refusal ordinary treatment or conditional suicide by omission.

I think that the rationale for an advance directive, even including the refusal of such things as ANH if one were in PVS, need not be suicidal and could indeed be a completely virtuous and generous decision to free caregivers from their duty to use their time and resources to provide one life-sustaining care. "Use your time, money and energy for other good things than keeping me going" is not necessarily suicidal, but can be a choice not to accept certain side effects of treatment, the costs, broadly conceived, of doing it. By the vocational standard of seriousness I suggested above, these side effects could reasonably be judged grave enough to justify foregoing them (Grisez, 1997, 218–225).

Of course, in our social environment, an advance directive to discontinue sub-stantial elements of life extending care such as ANH in conditions of great debility where prospects of significant recovery are lacking is likely to be based on a rejec-tion of living in that repugnant condition. That would be suicidal.

If cooperating with what was surely a decision of this kind were necessarily for-mal cooperation, then cooperators would be guilty of willing suicide, but respecting the social reality of the right to refuse treatment and its extension in advance di-rectives does not involve intending the evils of those who misuse their authority to decide about their health care. And within a pro-life community, the presumption should not exist that instructed people will be guilty of conditional suicide.

In short, I think advance directives may rightly be used to deal with many aspects of medical treatment and care anticipated for future times when one is not competent to make decisions. Such directives can be abused, and when they certainly are being abused, those who respect human dignity will cooperate as little as possible with them. But when thoughtfully made, they can introduce moral wisdom into situations in which family and friends may need some help.

## Notes

[1] See Boyle, 2002; I follow the classical magisterial formulation of Pius XII (Pius XII, 1958; more recent magisterial formulations include burdens as one of the things that can make a treatment "extraordinary" (John Paul II, 1995, paragraph 65).

[2] The prohibition against killing in Catholic teaching absolutely excludes only intentional killing of the innocent (John Paul II, 1995, paragraphs 55–66, 708–713).

[3] See Boyle [1981] for an exposition of Pius's position on a competent patient's refusal of medical treatment and a development of the thought in this paragraph.

# References

Boyle, J. (1980). Toward understanding the principle of double effect. *Ethics, 90*, 527–538.

Boyle, J. (1981). The patient/physician relationship. In D. McCarthy & A. Moraczewski, O.P., (Eds.), *Moral responsibility in prolonging life decisions* (pp. 80–94) St. Louis, Mo.: Pope John Center.

Boyle, J. (1995). A case for sometimes feeding patients in persistent vegetative state. In J. Keown, (Ed.), *Examining euthanasia: Legal, ethical and clinical perspectives* (pp. 189–198). Cambridge: Cambridge University Press.

Boyle, J. (2002). Limiting access to health care: A traditional roman catholic analysis. In H. T. Engelhardt, & M. Cherry (Eds.), *Allocating scarce medical resources: Roman catholic perspectives.* (pp. 77–95). Washington, DC: Georgetown University Press.

*Catechism of the Catholic Church: With Modifications from the Editio Typica.* (1997). (2nd Ed.). Washington, DC: U.S. Catholic Conference.

Grisez, G. (1997). *The Way of the Lord Jesus: Volume 3: Difficult Moral Questions.* Quincy II: Franciscan Press.

John Paul, II. (1995). Encyclical evangelium vitae. *Origins, 24*, 689–730.

John Paul, II. (2004). Care for patients in a "permanent" vegetative state. *Origins, 33*, 737, 739–740.

National Conference of Catholic Bishops. (1994). Ethical and religious directives for catholic health care services. *Origins, 24*, 450–461.

Pius, XII. (1958). The Prolongation of Life. *The Pope Speaks, 4*, 395–398.

# Chapter 7
# Understanding the Ethics of Artificially Providing Food and Water[1]

J. L. A. Garcia

In the short time since the late Pope John Paul II, now designated John Paul the Great, delivered his controversial March 2004 address on the morals of providing, withholding, declining, and discontinuing tube-feeding of persons in the condition often called "persistent vegetative state," a number of cases have arisen and interpretations offered of the papal statement, some of them seriously affecting patient care (John Paul II, 2004).[2] I wish here to offer a reading of the papal address that is more permissive than some in the options it preserves but, I think, faithful to the text and the wider tradition of Catholic moral reasoning in medical ethics. Much of my discussion will focus on the more general topic of assisted food and drink, rather than on the special issues raised by the condition misleadingly called "persistent vegetative state." I hasten to state that I am not a moral theologian, nor a theologian of any sort. Rather, I wish to offer reasons why the recent statement might, when construed within the larger context of Christian thought on ethical issues in medical practice, reasonably be judged consistent with that thought and with a position on responsible discretionary interventions with such patients that I find sensible and attractive. My proposed interpretation is offered as provisional, contingent on further exploration of these issues in the light of religious and secular moral inquiry. Moreover, we need to be sensitive to the fact that Pope John Paul II's March 2004 allocution is part of a tradition periodically developed and refined. Future doctrinal statements may clarify the tradition's implications, rendering obsolete some of today's judgments about what Catholic medical ethics requires, recommends, and permits.

I will proceed by first focusing on several key phrases in the papal statement, indicating for each how it might reasonably be seen to cohere with traditional Catholic ethical inquiry and moral common sense. In later sections, I relate the recent allocution to a number of recent, chiefly secular bioethical discussions of related issues, especially those of euthanasia and care of the incompetent.

J. L. A. Garcia
Boston College's Philosophy Dept.
Email: garciajl@bc.edu

C. Tollefsen (ed.), *Artificial Nutrition and Hydration:*
*The New Catholic Debate*, 123–139. © Springer 2008

## 7.1 Some Key Passages

Certain passages in the papal text have engendered much controversy and elicited criticism. We begin by examining some of them in this section. In the next, I offer interpretations of these texts to clarify them and suggest that many of the criticisms are misguided.

"The sick person in a vegetative state, awaiting recovery or a natural end, still has a right to basic health care (nutrition, hydration, cleanliness, warmth, etc. . . . [and] the right to appropriate rehabilitative care . . ." (John Paul II, 2004, no. 4).

"I should like particularly to underline how the administration of water and food, even when provided by artificial means, always represents a natural means of preserving life, not a medical act. Its use, furthermore, should be considered, in principle, ordinary and proportionate, and as such morally obligatory, insofar as it is seen to have attained its proper finality, which in the present case consists in providing nourishment to the patient and alleviation of his suffering" (John Paul II, 2004, no. 4).

"The evaluation of probabilities, founded on waning hopes for recovery when the vegetative state is prolonged beyond a year, cannot ethically justify the cessation or interruption of minimal care for the patient, including nutrition and hydration. Death by starvation or dehydration is, in fact, the only possible outcome as a result of their withdrawal. In this sense it ends up becoming, if done knowingly and willingly, true and proper euthanasia by omission" (John Paul II, 2004, no. 5).

Significantly, the pope quotes from his own 1995 encyclical *Evangelium Vitae*: "by euthanasia in the true and proper sense must be understood an action or omission which by its very nature and intention brings about death with the purpose of eliminating all pain" (John Paul II, 2004, no. 4).

"[N]o evaluation of costs can outweigh the value of the fundamental good which we are trying to protect, that of human life. Moreover, to admit that a decision regarding a man's life can be based on the external acknowledgment of its quality, is the same as acknowledging that increasing and decreasing levels of quality of life, and therefore of human dignity, can be attributed from an external perspective to any subject, thus introducing into social relations a discriminatory and eugenic principle" (John Paul II, 2004, no. 6).

## 7.2 Some Interpretive Suggestions

John Paul II's talk was given to an audience at a meeting organized by the Pontifical Academy for Life and the International Federation of Catholic Medical Associations. This audience, he could assume, would be familiar with Catholic tradition in medical ethics and would not need to have all relevant exceptions and context spelled out for them.[3]

(A) John Paul II describes providing artificial nutrition and hydration (hereafter, ANH) as both an "artificial means" and "a natural means of preserving life" (John Paul II, 2004, no. 4). This is apt to cause confusion. I think the passage is best read as indicating that ANH is an artificial way of doing a natural thing, as would be using a cell phone to call for help when trapped in a dangerous situation.

(B) The papal statement maintains that ANH is "not a medical act" (John Paul II, 2004, no. 4). Again, this is perplexing. It is, after all, a procedure performed by medical personnel using medical equipment in a medical setting. Some even prefer the term "medically assisted nutrition and hydration (MANH)" to "artificial nutrition and hydration," which I use here (Ford, 2004). In what respect is it, then, not a medical act? I suggest that the point is that providing nutrition is not a *healing*, *therapeutic*, or *disease-preventive* intervention. It has much, and arguably more, in common with what is commonly described as nursing care, such as keeping a patient clean and warm. My colleague, the eminent theologian Lisa Sowle Cahill compares ending ANH to removing respirators, and certainly there are important similarities.[4] Still, there are also differences that may matter. Breathing is naturally an involuntary motion, what followers of Aquinas sometimes call "an act of man" rather than a human action. In contrast, food and drink always enter our bodies through some agent's voluntary action, usually one's own.[5] That indicates an interesting and relevant way in which receiving ANH is closer to the natural and ordinary means of getting food and drink than is being on a respirator to natural and ordinary breathing. The fact that, in ANH, the means and setting of supplying food and drink are more technological does not suffice to make their provision narrowly *medicinal* rather than nutritive care.

Might it correctly be said that inserting and maintaining the feeding tube are medical acts, even if the feeding and watering are not?[6] Perhaps so, though this claim seems to rely on a dubious ontology of human action, according to which inserting and maintaining the tube that feeds the patient are somehow different actions from that of feeding her. If the feeding is not this, one wonders, what is it? In any case, I think the larger point is that little depends on this classificatory matter. Some people think it important because they reason that, since the language of "ordinary" and "extraordinary" means comes from medical ethics, if ANH is not medical, this distinction cannot apply to it, and thus no purchase is afforded the claim that it may sometimes be morally optional because extraordinary. This, however, is a *non sequitur*. If providing someone food and drink is excessively burdensome to her or others, it is not morally obligatory, regardless of whether the provision would be "a medical act" and whether or not it fits some restricted technical definition of an "extraordinary means."

(C) What of the address's explicit claims that ANH "should be considered, in principle, ordinary and proportionate, and as such morally obligatory" (John Paul II, 2004, no. 4)? Is withholding ANH, or even discontinuing it, then, never permitted?[7] That is not the only way to construe this claim, and I am not convinced that it is the best. First, compare the ordinary statement: What you borrow you ought in principle to return. Here, "in principle" means such return is a duty normally, considered in isolation and by itself (and *not* in the sense in which medieval thinkers used the term *in se*), in the abstract, in general. Adapting this use here could allow that ANH might still become extraordinary and disproportionate in some circumstances, and be licitly withheld or ended when it has. To be sure, the text admits of a stricter reading, supporting a more stringent rule. My suggestion, however, is that the new statement on ANH and persistent vegetative state (hereafter, PVS) be read as continuous with traditional Catholic thinking in healthcare ethics, so that considerations of proportion are relevant, and possibly dispositive, so long as the action is not of a type that in virtue of its nature can never be justified, is *malum in se*.[8] But should the recent papal statement be read precisely as branding withdrawal of (effective) ANH as *malum in se*? Again, it admits of such a reading, but at this point and waiting further clarification, I think that a looser reading is plausible, available, and open.

A year after the papal allocution, the National Catholic Bioethics Center in the USA announced its position that, "Food and water should be provided for all patients who suffer PVS unless it fails to sustain life or causes suffering," insisting that "[r]emoval of food and water is permissible only when they no longer attain the ends for which they are provided" (National Catholic Bioethics Center, 2005). However, it is not clear these two claims are themselves fully consistent. What of the case where ANH is not futile but attains its end of prolonging life, yet does so only by causing some burden so substantial as to be disproportionate? PVS may rule out the patient's feeling pain, but this is not the only form that a disproportionate burden may take.[9] The U.S. bishops conference has defined "disproportionate means" as measures "that in the patient's judgment do not offer a reasonable hope of benefit or entail an excessive burden, or impose excessive expense . . . "(United States Conference of Catholic Bishops, 2001). Of course, a PVS patient is in no position herself to judge "benefits" to be "reasonable" or a "burden . . . excessive." As Father Ford, of Australia's Chisholm Centre for Healthcare Ethics, affirms, such patients "are unconscious, unaware of themselves or their environment" and, because "awake but not conscious, they are unable to show their wishes" (Ford, 2004).[10] Nevertheless, it is not clear that morally ending (or withholding) ANH requires that the patient must make this judgment about, and at the time of, a proposed intervention. It may sometimes be enough that the burdens of initiating or continuing ANH go beyond some reasonable standard that the patient has endorsed previously and in the abstract, especially in a carefully drafted and reflectively informed advance directive.[11]

How might receiving ANH burden someone, specifically a PVS patient and her loved ones? Even if pain and discomfort are not issues for them, which can be questioned, and the expense is normally (but not always) modest, the toll in infections and other physiological complications may become so great that she could reasonably decline to continue despite the lethal result. Certainly, the costs to care-givers in time, stress, fatigue, can be so substantial it is morally permissible for them to decline further provision even in the face of the patient's likely death.

Does all this evacuate the papal injunction of its point, perhaps even content? Not at all. The point is that nourishing those unable to eat is valuable and justified quite independently of their "quality of life" or prospects for recovery. It should be the ordinary (here meaning typical, normal) thing to do, in no need of further justification. This is a truth it is important to assert in our time, when even some scholars in Christian medical ethics are wont to reject, sometimes deriding it as "vitalism," the claim that anyone's life is valuable in a way that warrants protection regardless of her disability, illness, etc. To the contrary, it is discontinuing such care that will always require special justification. That is not to say, however, that such justification can never be given, even when ANH prolongs life, and I do not read the papal text as excluding that possibility *tout court*. I return to both these points below.

(D) The 2004 address does allow that ANH might permissibly be withdrawn when it cannot achieve its "proper finality" (John Paul II, 2004, no. 4). Here, as the papal text makes clear, what is important is that the ends relevantly appropriate to the provision of food and water are nourishment and alleviating discomfort. Tubal delivery that cannot achieve this goal is futile and, at best, optional.

To be sure, there are problems in the offing at this point. Some insist that determinations of a form of treatment's futility themselves presuppose judgments of a patient's quality of life. Others distinguish such "qualitative futility" from a more empirical "physiological futility." To be sure, judgments that some type of care is futile are sometimes abused, used merely to camouflage the view that someone's life is not worth saving. Likewise, some determinations of futility do depend on evaluations of evidence. And efforts to eliminate any room for subjective judgment or assessment of evidence can have the effect of eliminating virtually any legitimate scope for considerations of futility.[12] For all that, it seems to me that a generally serviceable, attractive, reasonable, and objective account of futility can be given. Often we can make do with an account that holds a form of care for a type of illness to be futile when it has not achieved its more immediate physiological goals to any appreciable extent in the last hundred suitable patients within a region.[13]

Eating and drinking usually bring someone many advantages besides deliverance from hunger and thirst, of course, and they are often undertaken with these results chiefly in mind – pleasures of the palate, convivial enjoyment, a sense of one's welfare's being furthered, the interpersonal bonding that can be both manifested and cemented by sharing goods. The point is that the "proper finality" of eating is achieved in nourishment in the sense that this is its chief function, *telos*, point in nature/biology, though not necessarily the eating agent's foremost aim, motivating thought, or conscious and adopted personal objective. ANH that doesn't nourish and hydrate is futile treatment in the relevant sense, the one, as we saw, sometimes called physiologically futile.

(E) The pope stresses that "evaluation of probabilities . . . cannot ethically justify the cessation or interruption of minimal care for the patient, including [even artificial] nutrition and hydration" (John Paul II, 2004, no. 5). Must ANH, therefore, be started and maintained no matter what side effects are likely? I do not think the statement should be interpreted that narrowly. It is saying that evaluation of the probability of the patient's *recovery* is not necessary in order to make a determination of the medical and moral necessity of ANH. That is because such contingency would improperly suggest that the value of the patient's life hinges on its "quality", specifically, on the likelihood of her substantially recovering. That is false. The patient's life is valuable, just as graced with dignity as yours or mine, irrespective of her health. Nevertheless, the likelihood of both the nutritive *effectiveness* of ANH and of various untoward *side-effects* remains morally always relevant, and they should be continually monitored and assessed. I see nothing in the address that gainsays that.

(F) What of the text's excluding the possibility that our "evaluation of costs can outweigh the value of" human life, explicitly ruling out appeal to any "external acknowledgment of its quality . . . as acknowledging that increasing and decreasing levels of quality of life can be attributed from an external perspective to any subject," which latter step is described as "introducing into social relations a discriminatory and eugenic principle" (John Paul II, 2004, no. 5, 6)? There are several important and valid points made here. First, the sick and disabled have a serious (even equal) claim on our help. Second, only the person herself can determine at what point the burden of a form of intervention (whether or not it properly counts as a medical treatment) is too much for her. Presumably, a duly designated surrogate can try to determine and apply the patient's wishes, but none of us, separately or collectively, may substitute our judgment on what sort of life is worth living or saving.

That is not to say, of course, that the subject's own judgment is infallible, nor that it should always be the conclusive consideration. The papal statement is careful to indicate an asymmetry here. No one but the patient herself may properly authorize termination of ANH on the grounds that its relevant burdens (pain, discomfort, expense, intrusiveness, restrictions, inconvenience to and deprivation of others, esp., loved ones) have become disproportionate to its benefits to her (continued life). However, that does not mean that the patient must have the last word, no matter what it is. On the contrary, the pope does not rule out our relying on our own best judgment, rather than the patient's, in continuing ANH when we think that its benefits to her outweigh its burdens. This may seem unfair, and the thoughtless will complain that it is "cruel." However, it is not unfair, it is just asymmetrical, and the implicit guideline is that we need to take care to err, if sometimes we must, always on the side of life. It matters crucially *why* (the reasons for which) the medical team withdraws care and, in the case of the patient herself, it matters crucially why she asks that it be withdrawn.[14] As Kant, the father of modern autonomy, affirmed, not every preference is to be honored but only those wherein the agent acts free from pathological preferences and out of respect for her own and others' inherent dignity as persons.[15] The patient whose choice to end ANH or other care is a choice for death ought not to be abetted in this decision. In fact, this sort of death-wish, and its accompanying act of self-degradation, is one of the few places where the misused notion of "death without dignity" might find legitimate application. Privileging life over autonomy, as the pope here does, may displease some, but it is not an unreasonable prioritization. It is just the opposite, in fact, and plainly has nothing whatever of the character of genuine cruelty.

## 7.3 Withholding ANH in the Context of Euthanasia

We should now consider the papal claim that withdrawing ANH, thereby condemning the patient to "death by starvation or dehydration . . . as the only possible outcome . . . ends up becoming, if done knowingly and willingly, true and proper euthanasia by omission" (John Paul II, 2004, no. 4). This is quite strongly worded.

What is he saying here? Recall that John Paul II is careful in this address to quote from an earlier document where he sought to define euthanasia. There, he wrote that "by euthanasia in the true and proper sense must be understood an action or omission which by its very nature and intention brings about death with the purpose of eliminating all pain" (John Paul II, 1995, no. 65).

This makes it clear that, as this pope understands the term, intention is, along with result, a necessary condition for euthanasia. Intentions, of course, should be neither restricted nor imputed artificially or in a contrived way. We also need to remember that means count crucially. We will often need to ask, how does an agent reasonably mean/plan (that is, have it in mind) to get from her course of conduct $C$ to her envisioned resultant state $S1$ if not through the intermediary (means) step of result $S2$? As indicated in this account, it constitutes euthanasia, morally and in fact, when an agent's relevant course of conduct is self-restraining – that is, omissive – but nevertheless routed to its planned, targeted resultant state through the intermediary step of the patient's death. Again, it matters crucially *why* (the reasons for which) she withdraws care. And, in cases where the patient can voice her wishes, it matters crucially the reasons for which she asks that it be withdrawn. Those in PVS are, of course, unable, at the time of care, to voice or even have such preferences. However, they are not the only such patients, and it can be instructive for us to widen our scope and consider what people are saying about withdrawing life-sustaining treatment from, and even actively putting to death, neonates and other infants, especially, those with severe mental deficiencies.

Stephen Pinker, the prolific proponent and popularizer of so-called evolutionary psychology, wrote a controversial *New York Times* essay where he claimed,

> It seems obvious that we need a clear boundary to confer personhood on a human being and grant it a right to life. Otherwise, we approach a slippery slope that ends in the disposal of inconvenient people . . . [To recognize a right to life in all but only] members of our own species, Homo sapiens, . . . is simply chauvinism; a person of one race could just as easily say that people of another race have no right to life. No, the right to life must come, the moral philosophers say, from morally significant traits that we humans happen to possess: . . . having a unique sequence of experiences that defines us as individuals and connects us to other people, . . . an ability to reflect upon ourselves as a continuous locus of consciousness, to form and savor plans for the future, to dread death and to express the choice not to die. And there's the rub: our immature [human] neonates don't possess these traits any more than mice do (Pinker, 1997).[16]

I think Pinker's implicit comparison of the differences between White and Black people to those between the human and the subhuman racially offensive. Moreover, we do not "confer personhood on" people, nor "grant it" a moral right – note that term "it" – but rather recognize (acknowledge) their personhood and appreciate the rights that it grounds. Pinker notes that very few mothers who kill their one-day-old children (what he calls "neonaticide" as distinct from other "filicide" – that is, the killing of offspring) are tried, convicted, and imprisoned, and he infers from this that we empathize less with the very young victims. (He cites a study indicating that of 300 women charged with such crimes in USA and Britain, none spent more than one night in jail.) Pinker explains this differential concern by speculating that evolution has equipped us to feel a certain detachment until the child shows himself or herself capable of survival and thus a good investment of time and attention. However, Pinker himself notes that these baby-killers are usually very young, unmarried, alone, poor, and desperate. That suggests a different explanation for society's clemency: it may be rooted more in our greater sympathy for the killers than in our lesser sympathy for their victims. Finally, it is true neither that we "happen" to possess personal attributes nor that only those who in fact possess them, or possess them at a certain time of their lives, are persons. Rather, these qualities are most plausibly seen as defining personhood in that a human person is a being that by its natural inclination properly and naturally tends to develop them. When a human being is not yet at the stage where they have developed, or is at the appropriate stage but is such that some internal or external misfortunes have thwarted their development (or ended them) these facts do not deprive her of personhood, causally or conceptually. In fact, it only compounds her mischance, and literally adds insult to injury, if we take this disability as grounds further to deprive her even of social protections and personal respect.

Barbara Smoker, former president of Britain's National Secular Society, writes in a publication of the Council for Democratic and Secular Humanism,

> I strongly feel that it is cruel, and therefore immoral, to preserve a baby's life when there are such severe handicaps that chances of happiness are manifestly low. For life can, of course, be far worse than death.... [Acting] to starve seriously defective neonates to death – giving them only water, not milk... is certainly better than keeping them alive – but not as merciful as a quick, lethal injection.... Since we now have a social duty to limit our families, it is only sensible to limit them to those with a reasonable prospect of a normal human life.... A newborn baby has very limited awareness, no idea of any future, and no real stake in life, [whereas, in contrast,] an older child has become a real little person, with personal relationships, a sense of his or her own identity, and an idea of purpose – the very things that give human beings human rights and status (Smoker, 2003).

What we need to remember, in the face of Smoker's remarks, is that the principal way in which human life is valuable is rooted in the status and dignity of human personhood and does not derive from its usefulness for achieving happiness, having an idea of the future, and so on. Newborn babies are *already* and intrinsically persons, albeit little ones, and that status does not depend on what level of development they have achieved or will later achieve. H. Lagercrantz opines that

> it is wrong to ask if euthanasia of infants should be legal. It is better to retain respect for a personal life defined as a human individual with consciousness or the potential to become conscious. Having set his definition like this, he feels entitled to say, with regard deliberately to withholding resuscitation and treatment from certain severely encephalopathological children "with a very limited ability to develop a reasonable level of consciousness," that he "do[es] not regard this as euthanasia because the infant is not a conscious, or potentially conscious, person (Lagerkrantz, 2004).

Garret Keizer, an essayist who sometimes writes for the leftist American political magazine *The Nation*, believes he has uncovered what really motivates those who wish to protect life and medicine from the professional deformity that he and others call physician-assisted suicide. "The right talks about protecting life and tradition, but on some level . . . it is mostly interested in protecting pain. The first is theological: the belief that pain holds the meaning of life. . . . The second reason . . . is political: the belief that pain is fundamental to justice" (Keizer, 2005, p. 55).

Keizer offers little reason for imputing this odd idea to his adversaries, and it seems like mere unfairness. To be sure, many religious people remind us that suffering is not an unalloyed ill, and even biology shows its normal usefulness. They also know that deserved suffering is a necessary part of criminal justice, though most religious people in the West these days restrict the desirable types of suffering to the frustrations and restrictions of fines, incarceration, and, at most, relatively painless execution. More to the point, one would certainly have difficulty locating a reputable religious ethicist who thinks that medicine has a legitimate role in increasing, maximizing, prolonging, or enabling patients' pain. So, Keizer's claim is merely mean-spirited fancy.

Against what he imagines to be his adversaries' fixation on "protecting pain," Keizer wants to make his stand with liberal democracy. "PAS [physician-assisted suicide] rests on two principles that are central to a liberal society. The first is that we are owners of our own lives. . . . The second principle . . . is that we are collective owners of the culture we produce collectively. The debate over PAS is . . . about who owns the medicine. . . . And one thing more about the relevance of a Death with Dignity law to our democracy: we are free to try it out" (Keizer, 2005, p. 61).

Keizer's odd, left-libertarian view is rife with bad metaphors of ownership: owning our selves, owning medicine. It is unseemly for anyone to make so much depend on economic analogies, particularly someone ostensibly on the political Left. In any case, there is little substance to his reasoning, since it is difficult to make sense of his idea of owning ourselves – if this entitles us to kill ourselves does it not also entitle us to sell ourselves, even to give ourselves away to others in permanent bondage? But what sense can be made of self-alienation? And how could it be a matter of right? I cannot only destroy, sell, or give away my couch legally, I can also rip it up when I feel like it. And I can do any of these merely according to my passing caprice in the eyes of the law. Does Keizer's defense of PAS also bravely guarantee our moral right to deface, even mutilate, ourselves simply on the basis of a passing fancy? And should this entitlement also be legally encoded? If not, how and on what basis are these moral lines to be drawn? What sort of liberalism is it that could so casually commit itself to whimsical enslavement and mutilation? While John Locke spins in his grave, we would do well to remember this fact: those patients who exercise their Keizer-granted right to "try . . . out" PAS can never learn from their experiment's failure. It is doubtful that this is what Mill had in mind in talking of the marketplace of ideas.

Keizer also notices that many disabled people, and those who love them, are starting to notice the implications of condoning infanticide and mercy-killing for people who cannot take care of themselves, ambulate, and so on. Some do not like those implications, and are becoming more vocal in expressing their misgivings. "Groups like Not Dead Yet," he observes, "view any laws for assisted death as a threat to the very existence of the disabled. At least they claim to. After reading some of their literature, I suspect that what they see is not so much a threat as an insult. "Death with Dignity" becomes a loaded term in the presence of Life with Disability. Complaints about the "indignities" of terminal illness – loss of control over bathroom functions, complete lack of mobility – are naturally going to seem offensive to those who have struggled to assert their dignity under similar conditions (Keizer, 2005, p. 52)." Of course, the insult to the disabled in this nasty rhetoric is not merely perceived; it is inherent in much talk of "death with dignity."

John Robertson, an expert in issues of biomedical law, strains to justify his strange view that parents ought to be legally empowered to make care decisions because of what appears to constitute a plain conflict of interest. "Because parents (and other children) will bear the burdens of caring for the child with severe impairments, they should have the right to refuse resuscitation or treatment in . . . [severe] cases" (Robertson, 2004, p. 36). He continues, "If one lacks altogether the capacity for meaningful symbolic interaction, then one lacks the characteristics that make humans the object of moral duties. . . . [T]he mental disability in such cases is so extreme, so far from those cases in which children may be said to have valid interests in living, that they arguably do not threaten to harm the important values underlying the injunction against quality of life assessments in cases of disability" (Robertson, 2004, p. 37).

However, we should reject Robertson's claim that disabled babies lack relationship. Rather, they already *are* related to us, as somebody's son or daughter, grandchild, brother, sister, nephew, niece, etc. The question is how *we* respond to them – living up to (or failing in) these relationships. The same holds true for PVS patients.

This shift from concern to avoid burden on patients (especially, their pain), to avoiding burden on parents (who Robertson, oddly, wants to empower to make decisions precisely because of their conflict of interest), shows that not all the impetus for infanticide is really mercy-killing, killing from (supposed) mercy for the patient. It also raises the question: Which putative justification will come next? Avoiding *burden to society*? People who point out that the last century's chief advocates of euthanasia were in the Third Reich are nowadays denounced in high dudgeon. We have become accustomed to hearing that things are entirely different now, because the Reich supported euthanasia for the supposed good of society while the current trend appeals to individual autonomy and dignity. But as the rationale for infanticide shifts from avoiding the patient's pain to respecting her wishes, and now to enabling her potential care-givers to spare themselves expense and trouble, just how far are we from the Nazis' rationale for euthanasia?

Finally, consider Eduard Verhagen. Dr. Verhagen is clinical director of the Pediatric Clinic in the University Hospital of Groningen, a Dutch hospital whose physicians have a policy of sometimes performing mercy-killings on terminally ill newborns. In a recent NPR interview about his clinic's activities, he said, "[W]e felt that in these children the most humane course of action would be to allow the child to die, and even actively assist them in their death. . . . [F]rom a medical point of view, it is very important to be strict on the protection of life. And in extreme cases, the best way to protect life is to sometimes assist a little bit in death" (Montagne, 2004).

I will not belabor the nasty absurdities of humane infanticide and protecting life (whose? from what? one wonders) by helping "a little bit in death." One important lesson from reflecting on the disgraceful statements I have discussed, and it would be easy to multiply examples, is that we all need to learn to accept and value our common dependence, and recognize that there is no indignity in it, while it is manifestly degrading to judge some people to have lives worth neither saving nor living (see MacIntyre, 1999).

## 7.4 Withholding ANH: Intervention and its Discontents

Bearing those truths in mind can help us act more responsibly in making decisions about starting and discontinuing ANH. These decisions can get especially complicated with patients in PVS. We need to remember, for example, that diagnosis of PVS can be quite unreliable. Dr. Allen Counter reports that a 32-year old woman, whose doctors concluded that she "unresponsive to sensory stimuli, devoid of any intellectual function, and in a persistent vegetative state," consistently turned towards a music box playing in her room, and "began to smile and make sounds, as if she were enjoying" it (Counter, 2005, p. E1). He reports that he found himself "emotionally moved by her struggle for human definition through the single modality of hearing," that her doctors began responding to her more personally ("in some cases, holding her hand and trying to speak with her"), and that she continued to enjoy the music for some years. He concludes that "Her case was a reminder of how much we do not understand about the brain, and that even people in an apparent vegetative state may have ways of connecting to the world around them" (Counter, 2005, p. E2).

For all that, there can be reasons to withhold or discontinue ANH that are neither rooted in nor routed through the euthanasist's aim of "eliminating all [the patient's] pain" through her death. Dr. Muriel Gillick holds that feeding and watering by tube are "seldom warranted for patients in the final stage of dementia," because they have "few if any benefits and there is considerable potential for harm" (Gillick, 2000, p 309). She bases this conclusion on evidence that ANH has not been shown to lead to longer lives as compared with those who do not receive it, can cause diarrhea and nausea (so that neither nutrition nor hydration really results), and often leads to infections. In addition, demented patients frequently behave so as to pull out the tubes and need to be restrained, causing distress, fear, and diminished autonomy.[17] Moreover, ANH provides neither the felt satisfaction of eating nor the social interaction of being spoon-fed. Gillick's views have been controverted, and I certainly possess neither the knowledge nor the expertise to make a judgment.[18] My point is that the factors she cites are relevant, not excluded as such. Indeed, even the magnitude of financial costs to the patient and her family can be relevant, though this would be less true in a more just system of health care allocation.[19]

The US Catholic Bishops have stated that, while "there should be a presumption in favor of providing nutrition and hydration to all patients, including patients who require medically assisted nutrition and hydration," such intervention is morally required only when it "is of sufficient benefit to outweigh the burdens involved [by its provision] to the patient" (United States Conference of Catholic Bishops, 2001). Indeed, the burdens to others, e.g., relatives, can also legitimately be considered. Robert Orr reminds us that "For over a hundred years, traditional moral theology distinguished between ordinary and extraordinary means of saving life. Ordinary means were those that were not too painful or burdensome for the patient, were not too expensive, and had a reasonable chance of working" (Orr & Meilander, 2004). Other measures, involving "undue burden[,] were extraordinary and thus optional." The distinguished Protestant ethicist Gilbert Meilander rightly affirms that "we may refuse treatments that are either useless or excessively burdensome. In doing so, we choose not death but one among several lives open to us" (Orr & Meilander, 2004). His point is that sometimes, in withholding or discontinuing ANH, we act licitly because all that "we aim to dispense with [is] the treatment, not the life" (Orr & Meilander, 2004). In contrast, "if I decide not to treat because I think a person's life is useless, then I am taking aim at the not at the treatment but at the life" (Orr & Meilander, 2004). In this context, it is worth noting Father Ford's claim that "to prolong indefinitely the life of a patient in a permanent unconscious state does not seem to respect [her life's inherent] worth" and his alarming suggestion, cast as a question, that "subject[ing someone] to years of unconscious life sustained by MANH [medically assisted nutrition and hydration]" shows a "lack of respect for [that] patient's inherent dignity" (Ford, 2004).[20] I find it difficult to interpret this statement in a way that does not imply that it is such a life itself that is being targeted for termination because somehow unworthy, rather than the treatment that is being terminated *despite*, and *not because of*, the fact that doing so shortens the patient's life.

Even the idea of accepting some undesirable side effects of actions taken for good ends is complicated. We should not take too literally the metaphor of "weighing," Meilander sagely observes. "On what scale one 'weighs' benefits and burdens is a question almost impossible to answer. Even more doubtful is whether we can 'weigh' them for someone else. My own view is that when we make these decisions for ourselves, we are not weighing anything. We are deciding what sort of person we will be and what sort of life will be ours. We are making not a *discovery* but a *decision*" (Orr & Meilander, 2004). He cautions against being "too quick to assume that feeding tubes are 'treatment' rather than standard nursing care. . . . It is hard to see why such services as turning a patient regularly and giving alcohol rubs are standard nursing care while feeding is not" (Orr & Meilander, 2004). Meilander's trenchant discussion reminds us of our moral tradition's familiar insight that, as Father Ford nicely summarizes it, "Human life is a gift from God; it is a basic good of the person and not merely a means to other goods" (Orr & Meilander, 2004).

## 7.5 Conclusion

I have here tried to offer an interpretation and partial defense of the recent papal statement on ANH and PVS, placing this form of care and this condition in the larger context of recent discussions of care options and euthanasia, for other persons, especially newborns, who have diminished or unrealized capacities. Mine can be seen as a kind of middle view between those who deem the lives of some unworthy of prolongation, on one hand, and, on the other, an unconditional requirement to administer food and water artificially, provided only that it nourishes and hydrates, no matter how much doing so burdens the patient, and her family, and others. That is not to say I find the two alternatives equal. The more restrictive is a plausible, serious position, clearly consistent with responsible Christian thinking in medical ethics. In contrast, those who deem disabled lives unworthy of saving are spokespersons for what John Paul II has repeatedly and incisively labeled an "anti-life culture," even, perhaps better, an "anti-culture."[21] In some of my other writings on related topics, I have decried most of what today passes as biomedical ethics as a burlesque, a travesty. This field makes a mockery of its name, demeaning life (*bios*), perverting medicine from its proper ends, and suppressing that study and cultivation of the classical and Christian virtues (intellectual, moral, and theological) that constitutes genuine ethics.[22]

It used to be said that there is nothing deader than a dead pope, for the fate of his legacy and agenda rest in the hands of his successor. Surely, John Paul II is fortunate in having Benedict XVI, the former Cardinal Joseph Ratzinger, another formidable intellect and long trusted aide, succeed him. Still, I think the old saying exaggerated and less appropriate today, when mass media, a literate laity, and a Church that is highly educated in influential parts of world, also influence what is and is not kept alive and advanced through continuing discussions.

Early in 2004, as a particularly ugly attack on the most defenseless worked to its lethal conclusion in Florida over the admirable protests of human life's guardians, a local newspaper offered readers the following quotation from a prominent Catholic priest/theologian/"ethicist." "I think the best thing to do [with the March 2004 papal allocution] is ignore it."[23] His quoted remarks go on to deride "radical right-to-lifers" for trying to save the disabled victim from vicious starvation and to complain that such interventions as the papal address in March 2004 constitute "mischief-making in the Vatican." As quoted, such a voice gurgles from the blood-clogged sewers of our society's culture of death.

This sort of advice should guide indirectly, helping us see the necessity of doing the opposite of what it suggests. It makes it a pleasure for me to contribute to such a set of reflections on and interpretations of the pope's important recent call to protect innocent lives. I hope the essays collected here will stimulate further and continued appreciative reflections on the March 2004 allocution. In contrast, it is safe to ignore any moral theology that merely regurgitates for Catholic consumption the anti-life poisons it ingests from today's degraded and misnamed secular bioethics. While we continually revisit John Paul II's wisdom, let us disregard those forces of the culture of death who began swarming into Christian ethics decades ago from that theological revisionism Dietrich von Hildebrand presciently recognized as a Trojan Horse within God's City.[24]

# Notes

[1] I am indebted to Kevin McDonnell, David Solomon, and the staff of the 20th Notre Dame Medical Ethics Conference in London, March 2005, for inviting me to join them and for compiling for its participants most of the readings on which I draw here. I am grateful to Dr. Andrew McCarthy, of Britain's Linacre Centre for Healthcare Ethics, for additional bibliographical information. Paul Weithman and H. Tristram Engelhardt contributed illuminating discussions of these topics in London, while Christopher Tollefsen and Mark Sentesy kindly provided editorial help and suggestions. This discussion is dedicated to the memory of Pope John Paul II and Ms. Theresa Schiavo, during the final weeks of whose lives it was written, as one died under the artificial administration of nutrition and the other from what appears to me their unjustified and unjust removal. This essay originally appeared, with slight modifications, in *The Linacre Quarterly*. It is reprinted here with permission.

[2] This address was delivered to an International Congress sponsored by the Pontifical Academy for Life and the World Federation of Catholic Medical Associations.

[3] I have been told that the papal address was printed and delivered in Italian, and hastily translated by unknown staff. The doctrinal status of this papal statement is unclear. My colleague Father Himes casts doubt on the idea that it adds to "the official teaching of the Church," noting that it was "one papal speech to a special audience ... never promulgated to the universal church, nor were Episcopal conferences ever told to revise their local hospital directives" to bring them into conformity with the new teaching (Himes, 2005, p. 8). If I am correct that the statement does not really advance what Himes characterizes as "a novel position" but only builds on, clarifies, and develops a more familiar ethical stance, then the issue is not crucial. We should note, however, that in an age when even papal statements to local groups are quickly posted on the official Vatican website in several translations, the distinction between private remarks and what is universally announced may not longer be so sharp.

[4] Cahill claims the 2004 address is "marked by nonsequiturs and inconsistencies," immediately offering by way of illustration, "For one thing, it is hard to see how tube feeding can flatly be judged 'not a medical act' " (Cahill, 2005, p. 16).

[5] Watt makes a related point, while combating Father Ford's appeal to similarities between ANH and respirators: "Giving food and drink is ... part of non-medical, everyday care for many people, in a way that 'oxygenating' people is not (at least after birth). Infants and toddlers are routinely spoon-fed, as are disabled people of all ages. Tube-feeding is a low-tech extension of this kind of assistance: like the use of catheters, it is basic nursing care ... In any case, PVS patients often retain some ability to swallow, so that spoon-feeding would presumably need to replace the more convenient tube-feeding if that were withdrawn" (Watt, 2004). Contrast Ford, 2004: "Both [ANH and respirators], after all, use a medical procedure; in both cases death is the natural outcome unless ventilation or MANH is continued. Air and food are equally necessary for the maintenance of spontaneous life. If the ventilator may be ethically withdrawn, why not also MANH?"

[6] Australia's bishops suggested this in Australian Conference of Catholic Bishops, 2004.

[7] Never permitted, that is, except in the situation, described in the following paragraph, where ANH does not achieve its "proper finality" of nourishing the patient and therein palliating her discomfort.

[8] The consensus statement issued on this document by a 2004 Colloquium of the Canadian Catholic Bioethics Institute stresses this. The signatories affirm: "The papal speech needs to be understood in the context of the Catholic tradition. The words 'in principle' [in the passage cited] do not mean 'absolute' in the sense of 'exceptionless' but allow consideration of other duties that might apply." Their proposed gloss on the pope's statement seems to me well crafted and reasonable: "For unresponsive patients to whom ANH can be delivered without being in itself in conflict with other grave responsibilities or overly burdensome, costly or otherwise complicated, ANH should be considered ordinary and proportionate, and, as such, morally obligatory." They continue, "Treatments cannot be classified ahead of time as [inherently] ordinary or extraordinary.

Reference must be made to the wishes and values of the patient, his or her condition, and the availability of health care in the given context ... Extraordinary treatments are those that do involve excessive pain, expense, or other burdens." That determination is situation-based (Canadian Catholic Bioethics Institute, 2005, nos. 5, 7). In contrast, Cahill too confidently asserts that "the [papal] speech is not consistent with prior well-established teaching" (Cahill, 2005, p. 17). I seek a reading of the allocution that enables us to resist that extreme and implausible judgment.

[9] I say that PVS may "only rule out the patient's experiencing pain" because the matter has been controverted. Ford reports that the Congress to whom the pope made his March 2004 remarks also "heard evidence that some PVS patients had minimal consciousness, and that there was a possibility they could experience pain." As Ford notes, if and insofar as she can experience discomfort, the possibility is raised that ANH may be morally optional, discretionary, for a PVS patient on the grounds that the pain it causes her constitutes an excessive burden (Ford, 2004). My point is that, even in the absence of any possibility of pain, considerations of ANH's excessive and disproportionate burden may come into play because of the expense it runs the patient, her family, or others. In the last sections of this essay, I raise some ethical worries properly arising from recent theorists' emphasis on the social cost of sustaining life.

[10] A joint statement from the International Congress that the pope addressed describes a vegetative state as "a state of unresponsiveness, currently defined as a condition marked by: a state of vigilance, some alternation of sleep/wake cycles, absence of signs of awareness of self and surroundings, lack of behavioral responses to stimuli from the environment, [and] maintenance of autonomic and other brain functions" (International Congress, 2004, no. 1). What is important for our purposes is that characterizing the state in this way, chiefly by observed operations and (lack of) responses, leaves open the possibility that such a patient may experience pain in her inner life though she does not manifest typical pain behavior.

[11] To say this is not necessarily to endorse the confused counterfactuals that muddle much of today's debate. Plainly, there can be little sense to questions about what the PVS patient "would have wanted." This subjunctive indicates that the desire is conditioned on some situation. But what could it be? It is laughable to inquire whether she would or would not want (like?) being in PVS if she were in that condition (of a lack of awareness) and aware of it. Is it, then, that what matters is whether she would opt for a PVS life if described to her? But why should such a preference, essentially uninformed by any experience, be what counts?

[12] Bailey quotes a definition according to which "a particular medical treatment [is] futile if that treatment is incapable of accomplishing any of the specific goals of treatment" (Bailey, 2004, pp. 78–79). The "any" here is supposed to eliminate subjective judgments but, as Bailey sees, the account remains problematic. If the definition is meant to pick out only such treatments as *undeniably* have *no* chance of accomplishing any of their goals *to any extent*, it is so narrow as to have almost no application. If it is meant less restrictively, then room remains for judgments of likelihood and assessments of evidence. "[I]t will be rare, if not impossible, that the evidence will demonstrate that the particular intervention to be [sic] ineffective 100% of the time. What if statistical evidence shows that a particular intervention will succeed in achieving its goal 1% of the time? Strictly speaking, this intervention cannot then be labeled physiologically futile" (Bailey, 2004, p. 80).

[13] Here I draw on the account offered in Schneiderman and Capron, 2000, as amended by Bailey [2004] and further modified. I depart from Bailey, however, in holding that what matters for a (somatic) treatment's futility or efficacy is not whether it achieves its overall goal of effecting the patient's recovery but whether it makes the (causally) more immediate and smaller-scale physiological changes that it is hoped will contribute to her recovery.

[14] "Health care professionals may be confronted by patients who, with suicidal wishes, refuse ordinary life-sustaining care. Such patients must be treated with concern for their dignity and well-being. Health care professionals should do their best to protect the life and health of the patient while recognizing that there may be legal and professional limits to their ability to intervene ... [Nevertheless, a] Catholic health care professional should not cooperate in implementing a suicidal directive" (Canadian Catholic Bioethics Institute, 2005, nos. 12, 20).

[15] There are, thus, innocent mistakes that the patient might make, overestimating the suffering a treatment will cause (or its probability). Such failings in judging and reasoning are not normally culpable, even when born of emotional distortions, such as fear. Of course, controllable but uncontrolled fear, when excessive (or insufficient), can be a type of moral vice that can morally contaminate the judgments and decisions it shapes. Further, the patient may judge treatment too burdensome simply because she viciously despises a life of dependency, counting it as no benefit at all. Not all mistakes, then, about whether a mode of treatment's costs are too great, will be innocent. I am grateful to Professor Tollefsen for turning my attention to some of these points.

[16] Pinker does not specify which are "*the* moral philosophers" who root human rights in what "happens" to be true of us, but, unfortunately, there is no shortage of such thinkers who deny inherent dignity and restrict rights to those who have reached a certain level of development and accomplishment.

[17] The Canadian statement pertinently reminds us that "Some restraints may constitute an assault on human dignity . . . Restraints can also lead to complications such as pressure sores" (Canadian Catholic Bioethics Institute, 2005, no. 15).

[18] For the controversy, see Vollmann et al., (2000).

[19] The Canadian statement is again helpful on this point. "While recognizing that it is impossible to place monetary value on human life, the cost of treatment can be a morally relevant factor in health care decisions, especially if patients or their families have to bear the entire economic burden" (Canadian Catholic Bioethics Institute, 2005, no. 9).

[20] Similarly, Cahill is careful to distance herself from those who claim "that continued life would be a benefit no matter what its condition," and maintains that "it could reasonably be argued that 15 or more years of existence in a 'vegetative' state neither serves human dignity nor presents a fate that most reasonable people would obviously prefer to death" (Cahill, 2005, nos. 16, 17). I think it incoherent to deny that life is always a benefit to a human being, and can discern to disservice to human dignity in preserving a human life, in which dignity inheres as such and irrespective of the blocking of many normal capacities. On the contrary, to deem such a life as beneath preservation is to deny its inherent status. Whether many reasonable people would prefer death to a long life in PVS is morally irrelevant, since they may seek escape in death out of despair and incomprehension before the prospect of such a limited existence. Even reasonable people, of course, form some preferences from irrational parts of the self. Cahill suggests the chief issue is whether John Paul II 2004 allocution "settl[es] the question in favor of *always* using artificial nutrition." However, this threatens to mislead. She herself notes that Richard Doerflinger, a spokesperson for the U.S. bishops on pro-life matters, denies that the statement "declared an absolute moral obligation to provide assisted feeding in *all* cases." (Cahill, 2005, nos. 16, 17; emphases added) The papal statement explicitly repudiates ANH in cases where it is ineffective. Rather, the controversy today is, as Father Ford puts it, over the range of remaining discretion, what "wriggle room is now left for doctors in Catholic hospitals to continue to make decisions [about ANH for PVS patients] on a case-by-case basis." (Ford, 2004).

[21] For one such reference to our "anti-life culture," see John Paul II, 1981, no. 30; for "anti-culture," see his 1984 address to the Pontifical Council for Culture, cited in Dulles, 1999, p. 123.

[22] See Garcia [1999], and Garcia, 2007.

[23] I withhold the publication's title and date, and this theologian's name, out of charity and collegiality, to help protect his identity in hope that the remarks were misquoted and misconstrued.

[24] This deformation of Christian thought undergoes an important critique in John Paul II's 1993 encyclical, *Veritatis Splendor*. This revisionism is also sometimes called "proportionalism." As that name suggests, it stems from an exaggerated and distorted understanding of the appeal to proportionality in moral reasoning, one that misconceives its nature, limits, and preconditions. My point is that any position advanced as standing within Catholic moral theology must show that it has repudiated the errors of proportionalism and undergone the internal reform for which the late pope called. Without that, it can safely be ignored.

# References

Australian Conference of Catholic Bishops. (2004). Briefing Note on the Obligation to Provide Nutrition and Hydration. Available at: http://acbc.catholic.org.au/bbc/docmoral/2004090316.htm

Bailey, S. (2004). The concept of futility in health care decision making, *Nursing Ethics, 11*, 77–83.

Cahill, L. S. (2005). Catholicism, death and modern medicine. *America*, April, 25, 2005, 14–17.

Canadian Catholic Bioethics Institute. (2005). 'Reflection on artificial nutrition and hydration. *Consensus statement of a Canadian Catholic Bioethics Institute colloquium in Toronto*, available at: www.iacbweb.org/statement/2004.pdf

Counter, S. A. (2005). Music Stirred Her Damaged Brain. *Boston Globe, 267*(88), March 29, E1–E2.

Dulles, A. S. J. (1999). *The splendor of faith*. New York: Crossroad.

Ford, N. (2004). Impacts of papal teaching on vegetative patients in Catholic hospitals. *The Tablet* 1 May, 2004. Available at: onlineopinion.com.au/print.asp?article=2219.

Garcia, J. L. A. (1999). Beyond biophobic medical ethics. In K. Brinkmann (Ed.), *Proceedings of the twentieth world congress of philosophy, vol. 1: Ethics*, (pp. 179–188). Bowling Green: Philosophy Documentation Center.

Garcia, J. L. A. (2007). Reforming healthcare ethics. In D. Solomon, & M. Hogan (Eds.), *Medical Ethics at Notre Dame*. Notre Dame: University of Notre Dame Press, 1–23.

Garcia, J. L. A. (2007). Revisiting african american perspectives on biomedical ethics. In E. Pellegrino et al. (Eds.), *African american bioethics: culture, race, and identity*. Washington: Georgetown University Press, 1–23.

Gillick, M., M.D. (2000). Rethinking the role of tube feeding in patients with advanced dementia. *New England Journal of Medicine, 342*(3), 206–210.

Himes, K. O. F. M. (2005 June 6–13). To inspire and inform. *America, 7*–10.

International Congress on Life-Sustaining Treatments and Vegetative State: Scientific Advances and Ethical Dilemmas. (2004). Joint Statement. March 10–17, available at: http://www.life issues.net/writers/doc/doc_34vegetivestatejoint.html

John Paul II. (1995). Encyclical evangelium vitae. *Origins, 24*, 689–730.

John Paul II. (2004). Care for patients in a "permanent" vegetative state. *Origins, 33*, 737, 739–740.

Keizer, G. (2005). Life everlasting. *Harper's, 310* (1857), February, 53–61.

Lagerkrantz, H. (2004). Should euthanasia be legal? *Archives of disease in childhood. fetal and neonatal edition, 89*, F2.

MacIntyre, A. (1999). *Dependent rational animals: Why human beings need the virtues*. London: Duckworth.

Montagne, R. (2004). Interview with Dr. Eduard Verhagen. *National Public Radio*. December 1, 2004.

National Catholic Bioethics Center. (2005). The Case of Terry Schiavo. Press release of March 18, 2005, available at ncbcenter.org/press/05-03-18-Schiavo.html.

Orr, R., & Meilander, G. (2004). Ethics and life's ending: an exchange. *First Things, 145*, 31–37.

Pinker, S. (1997, November 2). Why they kill their newborns. *New York Times*, 1997.

Robertson, J. (2004). Extreme prematurity and parental rights after baby doe. *Hastings Center Report*, July/August, *34*, 32–39.

Schneiderman, L. J., Jecker, N. S., and Jonsen, A. (1996). Medical futility: Response to critiques. *Annals of Internal Medicine, 125*, 669–674.

Smoker, B. (2003). On advocating infant euthanasia (Op-Ed). *Free Inquiry, 24*, 17–18.

United States Conference of Catholic Bishops. (2001). *Ethical and religious directives for catholic health care services* (4th ed.). Washington: United States Conference of Catholic Bishops.

Vollmann, J. Burke, W. J., Kupfer, R. Y. Tessler, S. Friedel, D. M., Ozick, L. A., and Gillick, M. (2000) Correspondence. *New England Journal of Medicine, 342*(23), 1755–1756.

Watt, H. (2004, May 1). In defense of just being. *The Tablet*, available at: onlineopinion.com.au/print.asp?article=2294.

# Chapter 8
# The Ethics of Pope John Paul's Allocution on Care of the PVS Patient: A Response to J.L.A. Garcia

Peter J. Cataldo

The March 2004 allocution by Pope John Paul II, "Life-Sustaining Treatments And Vegetative State: Scientific Advances And Ethical Dilemmas," to a joint congress of the Pontifical Academy for Life and the International Federation of Catholic Medical Associations, has been studied as much if not more than any other of his pontificate (John Paul II, 2004). Views of the allocution's consistency with Catholic moral teaching and tradition I believe fall into three categories that I have characterized as the "major reversal view," the "contraction view," and the "recapitulation view" (Cataldo, 2004b, pp. 659–660). The major reversal view holds that the allocution runs contrary to Catholic moral tradition and teaching by indicating that the administration of nutrition and hydration is obligatory under all circumstances. The contraction view argues that the allocution has narrowed Catholic moral teaching and tradition in two ways: (1) by showing that the moral status of providing nutrition and hydration may be assessed apart from a patient's overall situation; and (2) by asserting that the moral obligation to provide nutrition and hydration to the patient in persistent vegetative state (PVS) may be judged strictly on the basis of whether nourishment can be attained. The recapitulation view finds that the allocution does not depart from Catholic moral teaching and tradition by reversing it or narrowing it, but rather in an application to the PVS patient recapitulates or summarizes the essential elements of Catholic teaching and tradition on the obligation to use food. My view is that the allocution recapitulates Catholic moral teaching and tradition.[1]

J.L.A. Garcia has written an insightful and helpful commentary on the pope's allocution, one which I find sees the allocution as recapitulating Catholic moral teaching and tradition on the use of food and water for the conservation of life. My response to Garcia's discussion will begin with one of the key passages from the allocution with which he begins. I will analyze and expand upon some of the important concepts and problems presented in Garcia's commentary on this text and other parts of the allocution by using the work of Saint Thomas Aquinas and some of the

Peter J. Cataldo
The National Catholic Bioethics Center, Philadelphia, PA
Email: Pjcataldo@comcast.net

C. Tollefsen (ed.), *Artificial Nutrition and Hydration:*
*The New Catholic Debate,* 141–161. © Springer 2008

classicists in Catholic moral theology. My response will examine such questions as: the ontology of action in the allocution; the moral status of the obligation to provide artificial nutrition and hydration (ANH); the meaning of an autonomous health care decision; and the personhood of the PVS patient. I will take the opportunity to argue a difference of opinion with Garcia on certain points, but will ultimately agree with Garcia's assessment of the pope's allocution.

## 8.1 The Allocution's Ontology of Action

The following text from the allocution is critically important as a focal point both for understanding the moral obligation regarding the use of food and water as a means of conserving human life, and for showing the historical continuity of the allocution with Catholic moral teaching and tradition in view of critiques of the allocution:

> I should like particularly to underline how the administration of water and food, even when provided by artificial means, always represents a *natural means* of preserving life, not a *medical act*. Its use, furthermore, should be considered, in principle, *ordinary* and *proportionate*, and as such morally obligatory, insofar as and until it is seen to have attained its proper finality, which in the present case consists in providing nourishment to the patient and alleviation of his suffering (John Paul II, 2004, no. 4).[2]

The allocution states, "the administration of water and food, even when provided by artificial means, always represents a *natural means* of preserving life, not a *medical act*." Garcia finds aspects of this claim to be confusing and perplexing. He also characterizes as "a dubious ontology of human action" that interpretation of the statement which holds that maintaining a feeding tube and feeding the patient are two different actions (Garcia, 2008, p. 129). The statement is thought to be confusing because it seems to describe ANH as both an artificial and a natural means of sustaining life. It is said to be perplexing because it asserts that ANH is not a medical act. Garcia adds that, "little depends on this classificatory matter" of whether or not ANH is a medical act (Garcia, 2008, p. 129). He makes these various points under two separate sections but they are closely interrelated and need to be analyzed together as I will do (see Garcia, 2008, sec. 7.2A and B).

All of the claims made in the paragraph of the pope's allocution under consideration, as I have stated, recapitulate the Catholic moral tradition on the obligation to use food for nourishment. In particular, the identification of the administration of food and water as a natural means of preserving life is consistent with the tradition. In order to appreciate better the historical continuity of the statement, it is important first to distinguish the concepts of means, action, use, and the end of preserving life in the Catholic moral tradition.

The intelligibility of a means is found in its relation to an end—a relation ordered by reason and tended toward by the will through an act of intention. This intelligibility is evident from the following description of intention by St. Thomas Aquinas: "The will does not order, but tends to something according to the order of reason. Consequently this word 'intention' indicates an act of the will, presupposing the act by which the reason orders something to the end" (*Summa Theologiae*, I–II, q. 12, 1, ad 3, trans. Benziger).[3] This "something" which reason orders to an end is the means. The means is also described by Aquinas as an intermediate term of movement of the will:

> Intention regards the end as a terminus of the movement of the will. Now a terminus of movement may be taken in two ways. First, the very last terminus, when the movement comes to a stop; this is the terminus of the whole movement. Secondly, some point midway, which is the beginning of one part of the movement, and the end or terminus of the other. Thus in the movement from A to C through B, C is the last terminus, while B is a terminus, but not the last. And intention can be both. Consequently though intention is always of the end, it need not be always of the last end (*Summa Theologiae*, I–II, q. 12, c. trans. Benziger).[4]

To summarize St. Thomas, the concept of means (*eorum*) refers to an intermediate terminus of the will, which is intermediate in relation to the ultimate terminus of the will as ordered by reason. This working definition accounts for the formality of means, but what sorts of items count as means? In his discussion of how choice is always of the means, Aquinas identifies means as being either actions or things: "For the means must needs be either an action; or a thing, with some action intervening whereby man either makes the thing which is the means, or puts it to some use" (*Summa Theologiae*, I–II, q. 13, a. 4, trans. Benziger).[5] It is important to note that a means is not equivalent to an act but includes things made and things put to a use. A human person acts by either moving some aspect of himself or herself with deliberate will, or by being moved in the same manner (see *Summa Theologiae*, I–II, q. 1, a. 1 and a. 3, trans. Benziger). Actions become means when they are placed by reason in an ordered relation to an end.

Another important concept for understanding the Pope's statement is the notion of "use" (*usus*) in the tradition. Aquinas states that the "use of a thing implies the application of that thing to an operation: hence the operation to which we apply a thing is called its use; thus the use of a horse is to ride, and the use of a stick is to strike" (*Summa Theologiae*, I–II, q. 16, a. 1, trans. Benziger).[6] Elsewhere, Aquinas describes use as applying an active principle (such as the appetitive power) to an action, and as "a movement of the appetite to something as directed to something else" (*Summa Theologiae*, I–II, q. 16, a. 2 c., trans. Benziger).[7] Hence, the use of a horse is intelligible as the movement of the appetite toward a horse for the purpose of riding, and the use of a stick as a similar movement of this active principle toward the stick for striking. The intelligibility of use also reveals that the concepts of use and means are mutually inclusive, as Aquinas indicates: "Use . . . implies the application of one thing to another. Now that which is applied to another is regarded in the light of means to an end; and consequently use always regards the means. For this reason things that are adapted to a certain end are said to be 'useful'; in fact their very usefulness is sometimes called use" (*Summa Theologiae*, I–II, q. 16, a. 3 c.).[8] For Aquinas, to make use of something by applying the active principle of the appetite to an action or a thing is to treat it as a means. In his concept of use, we are able to see how the distinct concepts of action, means, and use are interrelated.

An additional point about the concept of use, which is important for an understanding of the pope's statement, is that there are some things which may be per se useful but *per accidens* are not useful. The notion of per se usefulness is consistent with Aquinas' point made above that a use can be identified with certain things as when "their very usefulness is sometimes called use," and is also evident, for example, when it is said that the use of food and water is for the conservation of life. Aquinas addressed the concept of per se usefulness in his treatment of the topics of bodily mutilation and the nature of the good.[9] A part of the body is, in its very nature as a part, useful to the whole body. However, if a member of the body becomes diseased, it may become *per accidens* useless or harmful to the whole. The same part remains useful per se insofar as it is a part of a whole, but in the concrete circumstances this part may not be useful and may be removed, not insofar as it has a per se usefulness, but because accidentally it is a threat to the good of the whole (see *Summa Theologiae*, I–II q. 65, a.1 and ad 3, trans. Benziger).

In his reply to the objection that theologians should not consider circumstances of acts because circumstances can never formally qualify the good or evil of acts, Aquinas argues that circumstances do have this function inasmuch as that which is good is useful:

> Good directed to the end is said to be useful; and this implies some kind of relation: wherefore the Philosopher says (Ethic. i, 6) that "the good in the genus 'relation' is the useful." Now, in the genus "relation" a thing is denominated not only according to that which is inherent in the thing, but also according to that which is extrinsic to it: as may be seen in the expressions "right" and "left," "equal" and "unequal," and such like. Accordingly, since the goodness of acts consists in their utility to the end, nothing hinders their being called good or bad according to their proportion to extrinsic things that are adjacent to them (*Summa Theologiae*, I–II, q. 7, a 2 ad 1, trans. Benziger).[10]

From the perspective of the good as useful, it may be argued that food and water have inherent elements that are related to nourishment of the body which make food and water inherently useful (good) for the end of nourishment. Similarly, acts of feeding by the medical administration of food and water may be evaluated as good (useful) or bad according to whether the circumstances of this administration have a proper proportion to the end of nourishment.

The difference between the per se usefulness and *per accidens* ineffectiveness of food is clearly indicated by Aquinas in his *Summa Contra Gentiles*. In this work he attempts to prove that the use of food is not in itself sinful by showing that nutrition is the means by which the preservation of the body (the proper end of feeding) is achieved:

> Now, any action is performed in accord with reason when it is ordered in keeping with what befits its proper end. But the proper end of taking food is the preservation of the body by nutrition. So, whatever food can contribute to this end may be taken without sin. Therefore, the taking of food is not in itself a sin (*Summa Contra Gentiles*, III, II, 127, trans. Burke).[11]

Thus, in itself and per se food is useful. However, its sinful use is always due to some particular circumstance in which the use is *per accidens* contrary to its proper end:

Now, no food is by nature evil, for everything is good in its own nature, as we showed above. But a certain article of food may be bad for a certain person because it is incompatible with his bodily state of health. So, no taking of food is a sin in itself, by virtue of the type of thing that it is; but it can be a sin if in opposition to reason a person uses it in a manner contrary to his health (*Summa Contra Gentiles*, III, II, 127, 2, trans. Burke).[12]

We may identify three morally relevant factors in the act of feeding: (1) there is a proper end of feeding or taking food; (2) this end is nourishment insofar as a nourished body is a preserved one; (3) food by its nature and in itself a useful means for the preservation of life.

The classicists of the Catholic moral tradition following Aquinas all held the view that food is a per se useful means of preserving life.[13] Beginning with Francisco de Vitoria, O.P. (1492–1546), the classicists described the per se status of food, for example, as a "per se means ordered to the life of the animal and natural" (*alimentum enim per se medium ordinatum ad vitam animalis et naturale*);[14] as "per se intended" (*per se ordinata*) for the purpose of conserving life;[15] as "directed by nature for sustenance" (*ordinates a natura ad sustentationem*);[16] or as ordinary means "which nature has provided for the ordinary conservation of life" (*quae natura providit ad ordinariam vitae conservationem*).[17] The classicists also recognized that the use of certain foods in a given situation might be *per accidens* burdensome and not morally obligatory.[18] Moreover, foregoing food altogether for individuals who were seriously ill was also acknowledged as morally permissible (see Cronin, 1989, p. 35).

The apparent perplexity of the pope's statement that, "the administration of water and food, even when provided by artificial means, always represents a *natural means* of preserving life, not a *medical act*," is resolved when understood in light of the moral tradition pertaining to the distinctions between action, means, ends, and use. The administration of water and food for the patient is the act in question. This act represents a means in all three senses explained above. The act of administering food and water is a means *qua* action; it is a means *qua* application of one thing to another; and it is a means *qua* its per se use. First, the act of administering food and water is itself a means. The act itself is an intermediate end of the will ordered by reason toward the end of nourishment. Second, the act also includes use of a thing, an artificial feeding device. The will of the one acting is using the feeding device by applying one thing to another (the end of nourishment) and as such this use is considered a means. Third, this use of food and water is a per se useful or natural means toward the end of nourishment. It is in this sense that the medical administration of food and water is not a medical act. The act is not one of mere medical instrumentality, but rather has a utility that is per se and naturally suited for a particular end—the conservation of life.[19] Given the distinction between the three senses in which the act is a means, it is not the case that the pope proposes an act that has a contradictory nature, at once artificial and natural.[20] For the same reason, the pope does not imply that there are two different acts, a medical act of inserting and maintaining a feeding tube and another distinct act of feeding the patient (see Garcia, 2008, p. 129). Rather, there is one act that represents a means in three senses of the term.

Lisa Sowle Cahill asserts, similar to Garcia, that the pope's statement about ANH contains "nonsequiturs and inconsistencies" (Sowle Cahill Sowle, 2005, p. 16). However, given what the tradition says about means, ends, use, and the end of conserving life, this critique by Sowle Cahill is unwarranted. Moreover, contrary to what Cahill claims, a general permissibility to remove ventilators in comparison to ANH is not further evidence of a logically flawed statement about ANH by the pope, (see Sowle Cahill Sowle, 2005, p. 16). The use of mechanical ventilation and ANH are ethically evaluated by the same principle of ethically ordinary and extraordinary means. It is a *non sequitur* to conclude that the greater tolerance for withdrawal of mechanical ventilation in comparison to ANH is itself proof of inconsistent moral reasoning on the part of the pope. Any difference in the moral conclusions about mechanical ventilation and ANH is the function of a difference in burden as assessed by one and the same moral principle, not the result of a logical inconsistency. The fact that in any given case there may be more burden associated with prolonged mechanical ventilation than with ANH does not entail that the pope is inconsistent, but rather indicates that a greater burden associated with prolonged mechanical ventilation will be appropriately judged as ethically extraordinary when those burdens are present.

Garcia concludes his treatment of the pope's view of the medical status of ANH by arguing that "little depends on this classificatory matter" of designating ANH as either a medical act or not (Garcia, 2008, p. 129). He claims that the classification of ANH matters to those who hold that if the artificial administration of nutrition and hydration is classified as a medical act, then it will *wrongly* be considered ethically extraordinary. Garcia rightly points out that the single legitimate moral standard by which to evaluate the provision of food and water is whether it is excessively burdensome to the patient or others.[21] The moral standard for judging food and water as a life-sustaining means is not independent of, and different from, the standard used to judge other life-sustaining means. Clearly, for certain patients ANH is ethically extraordinary while for others it is ethically ordinary, and this is ultimately determined on whether it poses an excessive burden. However, it does not follow from this conclusion about the applicable standard for accessing ANH that little depends on how ANH is classified. In fact, properly assessing the burden of ANH in particular cases depends upon understanding that ANH is a natural means of conserving life in the sense discussed above. If, as the tradition holds, food and water is a per se useful or natural means for the end of conserving life, then this classification will be essential for correctly judging what set of circumstances counts as an excessive burden and what does not in any given case. Whether ANH in a given set of circumstances is an excessive burden is determined in relation to its status as a per se useful means of conserving life. It is important not to conflate the classification of ANH as a means with the moral standard by which obligations regarding this means are judged. If the two are conflated, then either the classification of ANH will be incorrectly regarded as being of little consequence, or ANH will be incorrectly viewed as having a moral standard *sui generis*.

## 8.2 Food and Water Considered "In Principle"

Garcia's view of the classification issue is related to how he interprets the sense in which ANH according to the pope "should be considered in principle ordinary and proportionate and as such morally obligatory." Garcia does not believe that "in principle" here has the sense of the term "*in se*" (in itself) as it was understood by medieval thinkers. Rather, he regards the use of "in principle" by the pope as having the meaning of a general or a prima facie moral duty:

> First, compare the ordinary statement: What you borrow you ought in principle to return. Here, 'in principle' means such return is a duty normally, considered in isolation and by itself (and *not* in the sense in which medieval thinkers used the term *in se*), in the abstract, in general. Adapting this use here could allow that ANH might still become extraordinary and disproportionate in some circumstances, and be licitly withheld or ended when it has (Garcia, 2008, p. 129).

Garcia's conclusion that the language of the pope's assertion does not exclude the possibility that ANH might legitimately be judged ethically extraordinary is correct. However, this need not be true only if "in principle" is interpreted as having the sense of a prima facie duty. In fact, both the prima facie duty and the judgment that ANH in a particular case is ethically extraordinary depend upon "in principle" being interpreted with the sense of the per se usefulness of food and water in the Catholic moral tradition (see Cataldo, 2004a, p. 529). There is a duty on the universal level to take food and water precisely because these are inherently ordered toward nourishment of the body. If food and water are per se useful for the conservation of life, then there is a moral obligation corresponding to this usefulness of food and water. This is not a moral generalization but a moral obligation integrally linked to a specific objective feature of the moral order.

Moreover, only if food and water have a per se usefulness with a corresponding obligation "in principle" to use them for the conservation of life can there be the possibility of a relative judgment in a particular case that ANH is not morally obligatory. A relative judgment about using ANH presumes that a proposed action in particular circumstances is being assessed relative to the per se usefulness of, and in principle obligation for, food and water. The question to be answered in the concrete case is whether the obligation in principle to use food and water must be fulfilled in the particular case; but this question is answered only if the per se usefulness of food and water is first recognized. Once it is known that the proper use of food and water for the human person cannot be achieved in the particular case, then the judgment can be made that accidentally they are not useful and therefore not obligatory.

If, in the particular case of a PVS patient, ANH will likely attain its "proper finality," the per se moral status of ANH may be fulfilled in the case and its administration is obligatory. However, the pope's reference to attainment of proper finality should not be equated with the issue of judging the futility of ANH, as Garcia seems to assume (see Garcia, 2008, p. 131). Garcia rightly calls attention to the distinction between "qualitative futility" and "physiological futility," and to how the judgment of futility is sometimes used in an attempt to medicalize a moral decision about the worth of a person's life (see Cataldo, 1991, pp. 3–4). He is also correct to explain the difficulty in quantifying the concept of futility to the point where it becomes useless as a guide for deciding when life-sustaining care and treatment are not morally obligatory (see Garcia, 2008, p. 131 and note 12). Garcia offers a definition of futility that is compatible with the pope's notion of the proper finality of ANH for the PVS patient. Garcia defines futile care: "a form of care for a type of illness ... [is] futile when it has not achieved its more immediate physiological goals to any appreciable extent in the last hundred suitable patients within a region" (Garcia, 2008, p. 131). This definition is helpful as a medical indication of futile care insofar as it takes account of the statistical success of a form of care generally for a class of patients with a particular type of illness. This standard of futility is an important factor in the attempt to arrive at a morally certain judgment that ANH either will or will not attain its proper finality in a particular case of PVS. However, the judgment that ANH is ethically extraordinary is not equivalent to a medical judgment of futility.

Futility is one factor among others in the consideration whether ANH has a reasonable hope of benefit for the patient and whether ANH will cause an excessive burden for the patient or for those who are responsible for the patient's care. The pope states that the proper end of ANH for the PVS patient "consists in providing nourishment to the patient and alleviation of his suffering." Several indicators of non-attainment of nourishment in addition to futility are implied in this statement. Nourishment for the PVS patient may be taken to mean the absorption of sufficient nutrients to maintain life at a constant level. Certainly, nourishment is not attained if the body cannot absorb nutrients. Here, and in the case of imminent and impending death, ANH is clearly futile. However, there are other situations in which ANH may cause excessive burdens due to serious complications, such as repeated aspiration-pneumonias, respiratory distress, or hydrocephalus.[22] Such effects from the administration of ANH may be sufficient to judge in certain cases that the ANH is ethically extraordinary irrespective of whether ANH nourishes the body. Moreover, the suffering that ANH alleviates for the PVS patient is the inability to swallow and/or ingest an adequate level of nutrients. If in an attempt to alleviate this suffering, ANH actually causes a serious complication for the patient, then presumably it may be judged that the ANH has not attained its proper finality of nourishment for the patient and is ethically extraordinary.

It is this way of failing to attain the proper finality of ANH to which The National Catholic Bioethics Center refers in a press release that Garcia criticizes. Garcia states that,

> A year after the papal allocution, the National Catholic Bioethics Center in the USA announced its position that, "Food and water should be provided for all patients who suffer PVS unless it fails to sustain life or causes suffering," insisting that "[r]emoval of food and water is permissible only when they no longer attain the ends for which they are provided." However, it is not clear these two claims are themselves fully consistent. What of the case where ANH is not futile but attains its end of prolonging life, yet does so only by causing some burden so substantial as to be disproportionate? (Garcia, 2008, p. 130)

The statement of the NCBC does not equate the non-attainment of the proper end of ANH with futility as Garcia assumes. Only by making this assumption can Garcia conclude that the claims of the NCBC seem to be inconsistent. Rather, the NCBC statement accurately reiterates the pope's dual condition of non-attainment by stating (1) that ANH is not morally obligatory if it fails "to sustain life" or if it "causes suffering," and (2) by referring to those situations in which food and water "no longer attain the *ends* for which they are provided [emphasis added]." Those cases which Garcia rightly points out involve substantial burdens despite adequate nourishment from ANH, are just the sort of cases that the NCBC statement recognizes as cases in which ANH causes suffering rather than ANH attaining the end of alleviating suffering (see Garcia, 2008, p. 130 and 137).

Indeed, it is conceivable that the excessive burden caused by the administration of ANH in some cases may prevent attainment of nourishment, as Garcia points out, for example, when ANH causes diarrhea and nausea (Garcia, 2008, p. 137). Thus, citing the failure of ANH to attain nourishment as being the primary condition under which ANH is not morally obligatory does not necessarily exclude recognition of excessive burden as a legitimate condition. Those cases notwithstanding in which ANH achieves adequate nourishment despite some excessive burden, the failure to attain nourishment may be directly due to a physiological cause, or it may be the result of some serious complication associated with ANH. In either case, nourishment as the proper finality of ANH is not attained and this makes ANH morally optional.

## 8.3 Prioritizing Life and Autonomy

Garcia comments on the moral weight of the subjective and objective elements present in the assessment of the obligation for ANH as that assessment is addressed by the following text of the papal allocution:

> First of all, no evaluation of costs can outweigh the value of the fundamental good which we are trying to protect, that of human life. Moreover, to admit that decisions regarding man's life can be based on the external acknowledgment of its quality, is the same as acknowledging that increasing and decreasing levels of quality of life, and therefore of human dignity, can be attributed from an external perspective to any subject, thus introducing into social relations a discriminatory and eugenic principle (John Paul II, 2004, no. 5).[23]

Garcia makes two important comments about this text: (1) "the sick and disabled have a serious (even equal) claim on our help"; and (2) "only the person herself can determine at what point the burden of a form of intervention . . . is too much for her" (Garcia, 2008, p. 132). It is the second point on which he focuses, analyzing both the subjective and objective dimensions of the judgment that ANH is excessively burdensome. Garcia begins his commentary by interpreting the pope's statement as emphasizing that no one can legitimately substitute his or her judgment for the patient's judgment about "what sort of life is worth living or saving" (Garcia, p. 132).

There are two difficulties with this interpretation. First, Garcia's language is inapt, because it is just this idea of a "sort of life" deemed worthy or unworthy that the pope is rejecting by arguing that life is a fundamental good which is not subject to a qualitative evaluation. Pope John Paul II is not claiming that there is a moral difference between a judgment made by a patient about what sort of life is worth living or saving and the same judgment made by the patient's caregivers. Rather, he is showing that this type of judgment—a quality of life judgment—is morally unacceptable whether made by the patient or not. Second, a quality of life approach to this issue is a relativistic one, and the thrust of the pope's statement is to show that consideration of ANH must be based upon absolute moral principles. This is evident both from the quoted text in which he identifies and applies the principle of the good of human life, and from the text that introduces his statement: "Considerations about the 'quality of life', often actually dictated by psychological, social and economic pressures, cannot take precedence over general principles" (John Paul II, 2004, no. 5).[24] Garcia seems to recognize this goal of the pope's statement by arguing that the patient's judgment is not infallible, that "it matters crucially" *why* a medical team or a patient choose as they do. Similarly, Garcia argues that the pope indicates an asymmetry existing between the objective guideline of always erring on the side of life on the one hand, and an autonomous decision on the other (see Garcia, 2008, p. 132).

However, Garcia interprets the pope's statement within the opposition of "privileging life over autonomy" (Garcia, 2008, p. 132). This opposition incorrectly assumes that a decision that privileges life against a patient's wishes is inconsistent with the autonomy of the patient. It is true that insofar as a patient's intention not to have ANH is not honored by the caregivers, they do not comply with that decision. However, not complying with a particular autonomous decision is not equivalent to overriding or ignoring the patient's autonomy. In fact, if the use of ANH is correctly judged to be ethically ordinary for a patient who otherwise does not want ANH, then a decision by the caregivers to give ANH represents a correction of the misuse of autonomy. Autonomy in the Catholic moral tradition is not identical with an independence or freedom from external limits and restrictions. Rather, rightful autonomy is the use of the will and conscience in fulfillment of the goods of human nature.[25] Proper authorization for ANH is solely reserved to the patient, but the subjective dimension of the decision does not exhaust the propriety of the authorization.[26] It is the patient who judges, but he or she must do so on the objective basis of the goods of human nature.[27] If the administration of ANH in any given case is an act in fulfillment of human good, then this act represents assistance and support to the patient's rightful autonomy (insofar as the caregivers are deciding on behalf of the patient), his or her misuse of autonomy through an advance directive notwithstanding. The moral asymmetry in the pope's statement is not an asymmetry of the privileging of life over autonomy, but rather an asymmetry between the privileging of life according to a rightful autonomy on the one hand, and a view of patient autonomy that is identified with the least restrictive decision of a patient on the other.

## 8.4 Intending, Being, and the Good of Human Life

Garcia comments on what for some is an alarming statement by Pope John Paul II regarding the withdrawal of ANH from the PVS patient, which is considered minimal care even when the prognosis is continued life beyond one year:

> Death by starvation or dehydration is, in fact, the only possible outcome as a result of their withdrawal. In this sense it ends up becoming, if done knowingly and willingly, true and proper euthanasia by omission (John Paul II, 2004, no. 4).[28]

Garcia rightly points out how the pope recognizes that both the result of the act and the intention of the agent must be evaluated in determining when the withdrawal of ANH is an act of euthanasia. However, Garcia seems to understand intention in euthanasia as being related solely to the result of the elimination of suffering:

> We also need to remember that means count crucially. We will often need to ask, how does an agent reasonably mean/plan (that is, have it in mind) to get from her course of conduct $C$ to her envisioned resultant state $S1$ if not through the intermediary (means) step of result $S2$? As indicated in this account, it constitutes euthanasia, morally and in fact, when an agent's relevant course of conduct is self-restraining–that is, omissive–but nevertheless routed to its planned, targeted resultant state through the intermediary step of the patient's death (Garcia, 2008, p. 135).

It is true that the means used to achieve a result are as morally significant as the end or result, and that using the death of the patient as the means to the intended end in the case of euthanasia determines the moral status of the act. However, as Aquinas' analysis of intention above showed, insofar as intention is the ordered movement of the will toward a terminus, both the means and end are intended—the former as an intermediate end and the latter as the last end. Not only is intention about both means and ends, but to intend an end is necessarily to intend the means to that end, as Aquinas argues, "a man intends at the same time both the proximate and the last end, as for example, the mixing of a medicine and the giving of health" (*Summa Theologiae*, I–II, q. 12, a. 3, c, trans. Benziger).[29] In any movement of the will toward means for the sake of an end, Aquinas argues further,

> the movement of the will to the end and its movement to the means are one and the same thing. For when I say: I wish to take medicine for the sake of health, I signify no more than one movement of my will. And this is because the end is the reason for willing the means (*Summa Theologiae*, I–II, q. 12, a. 4, c, trans. Benziger).[30]

Thus, given the intentionality of the means-ends relationship, Garcia's means-ends schema of the choice for euthanasia may be altered in the following way: the course of conduct (C) of withholding ANH is an intended means to reach the further intended intermediary step (S2) of the patient's death, in order to achieve the intended resultant state (S1) of elimination of suffering. Both the intermediary ends and the final end are intended in euthanasia.

It is important to note that although the elimination of suffering is one intended end in euthanasia, it cannot in reality be an object of choice by the will. This is because no such resultant state may exist as a positive state of affairs; but choice is only concerned with things in our power, not with "impossibles."[31] The intention to eliminate suffering by means of causing the death of a person is not a choice to eliminate suffering, because there is no subject to receive the chosen activity that purportedly results in the elimination of suffering. The elimination of suffering is a change in being that requires a subject of the change. A patient's suffering is eliminated or alleviated only if the patient receives a change from being in a state of suffering to no longer being in such a state. The unified, perduring presence of the subject from the beginning of the change to the end is necessary in order to conclude that a change has occurred in the subject. If the patient no longer exists as a result of withdrawing ANH, then there is no subject of a change in suffering, and as such, causing the death of the patient is not a chosen act of eliminating or alleviating suffering. Elimination of the patient is wrongly conflated with elimination of suffering because what is actually a recognition by others of the absence of the previous state of affairs is identified as a real state of affairs for the patient.

Garcia constructs an analogy between euthanasia by omission of ANH on the one hand, and euthanasia for neonates and infants with severe mental disabilities on the other, in order to highlight the reasons why euthanasia for PVS patients is not morally acceptable. Just as euthanasia is proposed by some as morally justifiable for patients in PVS due to their incapacity for rational self-aware consciousness, so also is euthanasia justified on the same basis in the case of life-sustaining care for certain cases of severely debilitated newborns and infants. For Garcia, the lack of capacity for rational self-aware consciousness and self-interests cannot be a justifying reason for euthanasia in both the neonate case and the PVS case.

Using the work of Stephen Pinker, Barbara Smoker, H. Lagercrantz, and John Robertson, Garcia provides several examples of the view that personhood is determined on the basis of having the capacity for rational self-aware consciousness and self-interests. It will be helpful to review some of relevant quotations from these authors on the subject given by Garcia. According to Pinker, "morally significant traits that we humans happen to possess [are]: . . . having a unique sequence of experiences that defines us as individuals and connects us to other people, . . . an ability to reflect upon ourselves as a continuous locus of consciousness, to form and savor plans for the future, to dread death and to express the choice not to die. And there's the rub: our immature [human] neonates don't possess these traits any more than mice do."[32] Smoker argues that "a newborn baby has very limited awareness, no idea of any future, and no real stake in life, [whereas, in contrast] an older child has become a real little person, with personal relationships, a sense of his or her own identity, and an idea of purpose—-the very things that give human beings human rights and status."[33] Robertson places emphasis on the capacity for interests by arguing that,

> If one lacks altogether the capacity for meaningful symbolic interaction, then one lacks the characteristics that make humans the object of moral duties. . . . [T]he mental disability in such cases is so extreme, so far from those cases in which children may be said to have valid interests in living, that they arguably do not threaten to harm the important values underlying the injunction against quality of life assessments in cases of disability."[34]

Garcia responds to the psychological account of personhood which these authors represent by arguing that the "human person is a being that by its natural inclination properly and naturally tends to develop" the characteristics of self-interest. He concludes pointing out that a lack of gestational development does not deprive an individual of personhood "causally or conceptually" (Garcia, 2008, p. 134).

This description of the kind of being that the human person is must be interpreted with an ontological meaning of capacity, not as a functional capacity for rational self-aware consciousness and self-interests. The difference that Garcia's statement makes compared to other arguments for personhood based on potential is his reference to the fact that the human person is a kind of being. Simply arguing that human persons have the potential to develop the relevant psychological qualities of personhood without reference to the ontological source or status of this potential is not sufficient to refute the proponents of psychological personhood. Such a potential is not an actual capacity for personal qualities in their view. Such a capacity is present only if there is evidence of self-interests or at least the physical substrate necessary for rational self-aware consciousness and the formation of self-interests.[35] According to the psychological view of personhood, the potential to develop the specific qualities that constitute personhood is in fact not to possess them at all, and to lack them is not to be a human person.

This view of possessing the capacity for personhood begs the question about the notion of capacity in several ways. It assumes that capacity is equivalent either to functionality (i.e., past or present operation of a specific quality) or to the physical substrate for functionality, and it assumes that the meaning of capacity cannot include what is potential because it assumes that whatever is potential cannot be real. These assumptions are themselves made on the basis of erroneous fundamental metaphysical assumptions about being itself. These authors rightly assume that action follows being but then draw the false conclusion that the activity of rational self-aware consciousness defines the being of the human person. This conclusion is not entailed by the principle that action follows being because it is wrongly assumed that the only principle of being is activity.

However, even though a human embryo or fetus does not demonstrate a function for rational self-aware consciousness and does not have a physical substrate sufficient for such consciousness, the embryo or fetus does have the capacity for qualities such as rational self-aware consciousness. These individuals have this capacity by virtue of the kind of being that they are; that is, because they exist according to a human nature which includes the operative potency to act or function with rational self-aware consciousness. Contrary to what the proponents of the psychological view assume, the particular activity of rational self-aware consciousness cannot be identified with existing as a human person for several metaphysical reasons: (1) the existence of a person is not the same as this self-conscious activity; (2) this activity may be interrupted but the individual's act of existing continues throughout such changes; (3) this activity, as an activity, is not in the category of what kind of thing the human person is; and (4) if the activity of rational self-aware consciousness were the same as a person's act of existing, then the act of existing would also be the potency for this activity to occur during its periods of interruption. However, this would be impossible because the person's the act of existing would be in act and in potency in the same respect at the same time.[36]

A proper metaphysical approach is needed to show that the activity of rational self-aware consciousness cannot be identical with human existing. Rather, this activity is a property of the kind of being the human is, i.e., it is a property of human nature. As a property of human nature, the fact that it is sometimes not actively present does not entail that it is completely absent; but rather, it is potentially present as a formal characteristic of human nature yet distinct from and not the same thing as human nature. Having the operative potency for the activity of rational self-aware consciousness makes a real difference among those individuals who count as persons. The way in which a human fetus, a doll, and a tree, for example, are said not to be rationally self-aware is really different. The latter two are not rationally self-aware because they are not rationally self-aware beings. The human fetus is not rationally self-aware because it is not actively self-conscious, but not because it is utterly incapable of being rationally self-aware. This metaphysical consideration of the issue indicates that what is real is not reducible to the actual, as the proponents of psychological personhood assume, but is composed of both the actual and the potential. The operative potency of the human embryo and fetus for rational self-aware consciousness is a real component of their existence, defines what sort of beings they are, and really differentiates them from other beings.[37]

The metaphysical basis of the personhood of the embryo, fetus, and PVS patient is presupposed in the following statement of Pope John Paul II:

> Today, the inviolable dignity of the person must be asserted more powerfully and consistently than ever! It is impossible to speak of a human being who is no longer a person or has yet to become one: personal dignity is a radical feature of each human being and disparity is neither acceptable nor justifiable (John Paul II, 2004b)!

With this statement, I believe that the Holy Father expressed a proposition that exceeds the content of previous magisterial statements on the issue. He is not saying as did the *Declaration on Procured Abortion* or the *Instruction on Respect for Human Life in Its Origin and on the Dignity of Procreation* that "the human being is to be respected and *treated as a person* from the moment of conception" (Congregation for the Doctrine of the Faith, 1987, I, 1). Rather, the pope has stated that every individual human being *is* a person, and as such the PVS patient is fully deserving of every consideration regarding the conservation of his or her life. That this statement is based on metaphysical grounds is indicated by his reference to the fact that any other alternative is "impossible" and that personal dignity is "a radical feature" of the human being. To exist as an individual human—either as an embryo, fetus, or in a PVS—is to possess the qualities of human dignity in a radical way, as rooted in the potency-act structure of existence.

The good of life of every person is a good in its own right apart from other higher goods which it may serve. Gilbert Meilander's and Fr. Norman Ford's comments on the relation of ANH toward the good of life are helpfully presented by Garcia:

> He [Meilander] cautions against being "too quick to assume that feeding tubes are 'treatment' rather than standard nursing care. . . . It is hard to see why such services as turning a patient regularly and giving alcohol rubs are standard nursing care while feeding is not." Meilander's trenchant discussion reminds us of our moral tradition's familiar insight that, as Father Ford nicely summarizes it, "Human life is a gift from God; it is a basic good of the person and not merely a means to other goods" (Garcia, 2008, p. 138).[38]

This recognition of the good of life as being an ultimate end in its own order of means-ends relations is critically important for evaluating the claim that ANH is not obligatory for the PVS patient because a person in this condition can no longer direct the use food and water toward the spiritual purposes of life.[39] As we have noted, the Holy Father states in the allocution that the "proper finality" of food and water for the PVS patient is nourishment. Nourishment is stated as the proper end despite the fact that the PVS patient does not have the functional capacity to pursue the higher spiritual good of life to which the limited good of physical life is subordinate. The

Holy Father's statement is consistent with the Catholic moral tradition because the tradition recognized that the good of life is the type of good which is both ultimate in its own order and is a means to a higher good.[40]

For example, as was shown earlier, Aquinas recognized that a means is itself an intermediate end, and that the good of the body is such an end. "Just as the body is ordained to the soul, as its end," he points out, "so are external goods ordained to the body itself" (*Summa Theologiae*, I–II, q. 2. a. 5, ad 1, trans. Benziger).[41] For St. Thomas, health is also an example of an "end in one operation" that is also "ordered to something as an end" (*Summa Theologiae*, I–II, q. 13. a. 3, c, trans. Benziger).[42] Health is the end of the physician's work but with respect to the good of the soul, health is to be regarded as a means. When it comes to the issue of taking less food, reason, St. Thomas states, may "direct this to the avoidance of spiritual evils and the pursuit of spiritual goods. Yet reason does not retrench so much from one's food as to refuse nature its necessary support" (*Summa Theologiae*, I–II, q. 147, a.1, ad 2, trans. Benziger).[43] What these quotations from St. Thomas show, as well as those presented earlier from the classicists, is an inherent structure of bodily life as an ultimate end in its own order.

The status of the life of the body as an end to which the means of food and water are ordered does not cease to exist as an end when the functional capacity to direct bodily health to the spiritual good is lost. In this circumstance, food and water as means are still ordered to the good of the body insofar as this good remains the natural end of those means. To say that nourishment is the proper finality of administering food and water to the PVS patient, is to recognize the inherent and permanent ordering of bodily life as an ultimate end relative to food and water as means for the fulfillment of that end. However, whether the administration of food and water for conserving life is actually warranted in any particular PVS case depends upon the individual circumstances. The fact that bodily life is an ultimate end in its own order does not preclude the possibility that in any given case nutrition and hydration will not be a proportionate means of conserving life as was discussed earlier.[44]

## 8.5 Conclusion

Garcia concludes his discussion by stating that his interpretation

> can be seen as a kind of middle view between those who deem the lives of some unworthy of prolongation, on one hand, and, on the other, an unconditional requirement to administer food and water artificially, provided only that it nourishes and hydrates, no matter how much doing so burdens the patient, and her family, and others. That is not to say I find the two alternatives equal (Garcia, 2008, p. 139).

This middle view is in fact the view of the allocution. It is a view that is consistent with Catholic moral teaching and tradition. Garcia's interpretation is confirmed by an examination of the Catholic moral tradition on the status of food and water as per se useful means of conserving human life, which in the particular case and *per accidens* may not be useful. This tradition allows the pope to assert consistently on the one hand that ANH is a natural means and in principle is proportionate and morally obligatory, and on the other indicate that ANH is not morally obligatory in each and every case.

The pope shows that the condition of PVS is not by itself reason to withdraw or withhold ANH as if the PVS patient is somehow less than human, or because the patient lacks the functional capacity to direct the good of life toward his or her spiritual good. The Holy Father has affirmed in his allocution that the intrinsic good of life is ultimate in its own order and as such is unaffected by the condition of PVS. This fact entails that PVS does not of itself make the administration of ANH an ethically extraordinary means of conserving human life. There may be any number of reasons why ANH is ethically extraordinary for a particular PVS patient, but being permanently unconscious is not one of them. Jorge Garcia has shown that this is a fundamental truth brought to light by Pope John Paul's allocution.

## Notes

[1] See Cataldo (2004b, pp. 513–536); see also O'Brien et al. (2004, pp. 660–662).

[2] The pope originally delivered this text in Italian: "In particolare, vorrei sottolineare come la somministrazione di acqua e cibo, anche quando avvenisse per vie artificiali, rappresenti sempre un mezzo naturale di conservazione della vita, non un atto medico. Il suo uso pertanto sarà da considerarsi, in linea di principio, ordinario e proporzionato, e come tale moralmente obbligatorio, nella misura in cui e fino a quando esso dimostra di raggiungere la sua finalità propria, che nella fattispecie consiste nel procurare nutrimento al paziente e lenimento delle sofferenze" [On-line] (Available: http://www.vatican.va/holy_father/john_paul_ii/speeches/2004/march/documents/hf_ jp-ii_spe_20040320_congress-fiamc_it.html).

[3] "Voluntas quidem non ordinat, sed tamen in aliquid tendit secundum ordinem rationis. Unde hoc nomen intentio nominat actum voluntatis, praesupposita ordinatione rationis ordinantis aliquid in finem."

[4] "Intentio respicit finem secundum quod est terminus motus voluntatis. In motu autem potest accipi terminus dupliciter, uno modo, ipse terminus ultimus, in quo quiescitur, qui est terminus totius motus; alio modo, aliquod medium, quod est principium unius partis motus, et finis vel terminus alterius. Sicut in motu quo itur de a in c per b, c est terminus ultimus, b autem est terminus, sed non ultimus. Et utriusque potest esse intentio. Unde etsi semper sit finis, non tamen oportet quod semper sit ultimi finis."

[5] "Quia necesse est ut id quod est ad finem, vel sit actio; vel res aliqua, interveniente aliqua actione, per quam facit id quod est ad finem, vel utitur eo."

[6] "Usus rei alicuius importat applicationem rei illius ad aliquam operationem, unde et operatio ad quam applicamus rem aliquam, dicitur usus eius; sicut equitare est usus equi, et percutere est usus baculi."

[7] "... motum appetitus ad aliquid in ordine ad alterum. ..."; see also *Summa Theologiae*, I–II, q. 16, a. 2, ad 1).

[8] "Uti ... est, importat applicationem alicuius ad aliquid. Quod autem applicatur ad aliud, se habet in ratione eius quod est ad finem. Et ideo uti semper est eius quod est ad finem. Propter quod et ea quae sunt ad finem accommoda, utilia dicuntur; et ipsa utilitas interdum usus nominatur."

[9] See *Summa Theologiae*, I–II, q. 65, a 1 c. and Ibid., I–II, q. 7, a 2 ad 1; see also Cataldo (2004a, pp. 522–523).

[10] "Bonum ordinatum ad finem dicitur utile, quod importat relationem quandam, unde philosophus dicit, in I Ethic., quod *in ad aliquid bonum est utile.* In his autem quae ad aliquid dicuntur, denominatur aliquid non solum ab eo quod inest, sed etiam ab eo quod extrinsecus adiacet, ut patet in dextro et sinistro, aequali et inaequali, et similibus. Et ideo, cum bonitas actuum sit inquantum sunt utiles ad finem, nihil prohibet eos bonos vel malos dici secundum proportionem ad aliqua quae exterius adiacent."

[11] "Fit autem unumquodque secundum rationem quando ordinatur secundum quod congruit debito fini. Finis autem debitus sumptionis ciborum est conservatio corporis per nutrimentum. Quicumque igitur cibus hoc facere potest, absque peccato potest sumi. Nullius igitur cibi sumptio secundum se est peccatum."

[12] "Adhuc. Nullius rei usus secundum se malus est nisi res ipsa secundum se mala sit. Nullus autem cibus secundum naturam malus est: quia omnis res secundum suam naturam bona est, ut supra ostensum est. Potest autem aliquis cibus esse alicui malus inquantum contrariatur salubritati ipsius secundum corpus. Nullius igitur cibi sumptio, secundum quod est talis res, est peccatum secundum se: sed potest esse peccatum si praeter rationem aliquis ipso utatur contra suam salutem." These and other passages in the *Summa Contra Gentiles* are also relevant to the issue of the subordination of physical life to the spiritual good of the human person; see Ibid., III, II, 123, 1 and 2; 127, 4, and 7; see this issue discussed below in text.

[13] See Cataldo (2004a, pp. 524–525); see also Cronin (1989, pp. 33–66). English translation of the classicists used herein is from Cronin [1989].

[14] Cronin (1989, p. 35) quoting de Vitoria, O.P, *Relectio de temperantia*, no. 1.

[15] Ibid.

[16] Cronin (1989, p. 43) quoting Thomas Sanchez, S.J., *Consilia seu Opuscula Moralia*, Tom. II, Lib. V, Chap. 1, dub. 3.

[17] Cronin (1989, p. 52) quoting Juan de Lugo, S.J., *Disputationes scholasticae et morales*, VI, *De iustitia et iure*, disp. X, sec I, no. 2.

[18] See Cataldo (2004a, p. 525) and Cronin (1989, p. 42).

[19] Garcia is correct that the act is not a healing one, but contrary to Garcia, the act is therapeutic insofar as it ameliorates the inability of the patient to ingest food by mouth.

[20] Garcia (2008, pp. 128–129) proffers the restatement that "ANH is an artificial way of doing a natural thing" as a better expression of the pope's point. However, this suggestion may equally be interpreted as asserting contradictory qualities of one and the same act.

[21] The judgment of excessive burden for medically administered nutrition and hydration frequently includes the judgment that the means have no reasonable hope of achieving their end.

[22] See (O'Brien et al. 2004, p. 507); see also Murphy and Lipman (2003, pp. 1351–1353) and Gilllick (2000, pp. 206–210).

[23] "Innanzitutto, nessuna valutazione di costi può prevalere sul valore del fondamentale bene che si cerca di proteggere, la vita umana. Inoltre, ammettere che si possa decidere della vita dell'uomo sulla base di un riconoscimento dall'esterno della sua qualità, equivale a riconoscere che a qualsiasi soggetto possano essere attribuiti dall'esterno livelli crescenti o decrescenti di qualità della vita e quindi di dignità umana, introducendo un principio discriminatorio ed eugenetico nelle relazioni sociali" [On-line] (Available: http://www.vatican.va/holy_father/john_paul_ii/speeches/2004/march/documents/hf_jp-ii_spe_20040320_congress-fiamc_it.html).

[24] "Su tale riferimento generale non possono prevalere considerazioni circa la "qualità della vita", spesso dettate in realtà da pressioni di carattere psicologico, sociale ed economico."

[25] See Pope John Paul II on the notion of "rightful autonomy": "At the heart of the moral life we thus find the principle of a 'rightful autonomy' of man, the personal subject of his actions. . . . The rightful autonomy of the practical reason means that man possesses in himself his own law, received from the Creator. Nevertheless, *the autonomy of reason cannot mean that reason itself creates values and moral norms*" (Pope John Paul II, 1995, no. 40).

[26] For the view that the moral justification of decisions to withdraw or withhold life-sustaining care and treatment is identified with a subjective notion of proper or valid authorization see Beauchamp and Childress (2001, p. 143); see also Beauchamp (2004, pp. 118–129).

[27] The language of directives 57 and 58 in USCCB (2001) leaves the relationship between the objective and subjective dimensions of the patient's judgment about what is ethically ordinary and extraordinary unclear insofar as all the elements of these judgments are unqualifiedly said to be "in the patient's judgment"; see Cataldo (1995, pp. 173–174).

[28] "La morte per fame e per sete, infatti, è l'unico risultato possibile in seguito alla loro sospensione. In tal senso essa finisce per configurarsi, se consapevolmente e deliberatamente effettuata, come una vera e propria eutanasia per omissione."

[29] "Simul autem intendit aliquis et finem proximum, et ultimum; sicut confectionem medicinae, et sanitatem."

[30] "Et sic unus et idem subiecto motus voluntatis est tendens ad finem, et in id quod est ad finem. Cum enim dico, volo medicinam propter sanitatem, non designo nisi unum motum voluntatis. Cuius ratio est quia finis ratio est volendi ea quae sunt ad finem."

[31] See Aristotle, *Nicomachean Ethics*, III, 2, 1111b 20–25; see also St. Thomas Aquinas, *Commentary on Aristotle's* Nicomachean Ethics, Lec. V, 443–445. The fact that the elimination of suffering is not open to choice in euthanasia would seem to make the corresponding will act a wish and not an intention.

[32] Garcia (2007, pp. 11–12) quoting Pinker [1997].

[33] Garcia (2007, p. 13) quoting Smoker (2003, pp. 17–18).

[34] Garcia (2007, p. 16) quoting Robertston (2004, p. 37).

[35] In addition to the authors Garcia cites see also Singer (1993, pp. 182–190): "The fact that a being is a human being, in the sense of a member of the species Homo sapiens, is not relevant to the wrongness of killing it; it is, rather, characteristics like rationality, autonomy, and self-consciousness that make a difference. Infants lack these characteristics. Killing them, therefore, cannot be equated with killing normal human beings, or any other self-conscious beings. This conclusion is not limited to infants who, because of irreversible intellectual disabilities, will never be rational, self-conscious beings. . . . A right to life to a being apply only if there is some awareness of oneself as a being existing over time, or a continuing mental self. Nor can respect for autonomy apply where there is no capacity for autonomy. The remaining principles [to use] are utilitarian. Hence the quality of life that the infant can be expected to have is important. . . . If disabled newborn infants were not regarded as having a right to life until, say, a week or a month after birth it would allow parents, in consultation with their doctors, to choose on the basis of far greater knowledge of the infant's condition than is possible before birth."

[36] See Klubertanz (1963, pp. 126–127); see also Clarke, (2002, pp. 118–121). Both authors deftly demonstrate the dependence of specific activities on corresponding active or operational potency within the nature of an individual.

[37] The invalidity of the assumption that what is real about the human person is reducible to the actual is also evident from the incorrect inference that because the operative potency for rational self-aware consciousness is not empirically *discoverable* precisely as potency, that it is in no sense real. Even though this operative potency is not empirically discoverable precisely as a potency, it is nevertheless metaphysically known as something that must be real. To assume that it is not real is to conflate the empirical mode of discovery with what counts as real.

[38] Garcia is quoting Meilander [2004] and no. Ford [2004].

[39] For examples of this spiritual purposes view of the moral obligation to use ANH see: O'Rourke (1988, p. 32; 1989, p. 188; 1991, p. 14; 1992, p. 24; 1999, pp. 3–4; 2002, p. 599; 2003, pp. 16–17), Ashley and O'Rourke (1997, p. 425), McCormick (1978, pp. 34 and 36; 1981a, pp. 345–348; 1981b, pp. 367–368; 1992, p. 211), Panicola (2001a, pp. 18, 21–22; 2001b, pp. 28–30, 33), Hamel and Panicola (2004, p. 7) and Kopfensteiner (2000, p. 21); for a secular analogue of this position see Shannon and Walter (1988, pp. 633–634, 644–645).

[40] See Cataldo (2004a, pp. 530–535); and see Ashley (1994, pp. 79, 81, 84, 86, 88, 92) for a very helpful examination of the concept of intermediate ends that are ultimate in their own order.

[41] "sicut corpus ordinatur ad animam sicut ad finem, ita bona exteriora ad ipsum corpus."

[42] "ita etiam contingit id quod est in una operatione ut finis, ordinari ad aliquid ut ad finem."

[43] "ratio recta hoc ordinat ad spiritualia mala vitanda et bona prosequenda. Non tamen ratio recta tantum de cibo subtrahit ut natura conservari non possit."

[44] For a sample of other views in opposition to the spiritual purposes argument concerning the use of ANH see: May (1990, 1997, 1998, 1999, 2000, pp. 255–257), Grisez, (1990) and Committee for Pro-Life Activities, National Conference of Catholic Bishops (1992, pp. 6–7); and R.M. Doerflinger [2004].

# References

Aquinas, Thomas. Saint. (1993) *Commentary on Aristotle's Nicomachean Ethics*. Notre Dame, IN: Dumb Ox Books.

Aquinas, Thomas, Saint. (1993) *Summa Theologiae* [On-line]. Available: http://www.thomas instituut.org/thomasinstituut/scripts/index.htm.

Ashley, B. M. O.P., (1994). What is the end of the human person? The vision of god and integral human fulfilment. In L. Gormally, (Ed.). *Moral truth and moral tradition* (pp. 68–96) Dublin: Four Courts Press.

Ashley, B. M. O.P. & O'Rourke, K. D. O.P. (1997). *Health care ethics: A theological analysis*, (4th ed.) Washington, DC: Georgetown University Press.

Beauchamp, T. L. (2004). When hastened death is neither killing nor letting die. In T. E. Quill, & M. P. Battin (Eds.), *Physician-assisted Dying: The case for palliative care & patient choice* (pp. 118–129). Baltimore: The Johns Hopkins University Press.

Beauchamp, T. L., & Childress, J. F. (2001). *Principles of biomedical ethics*, (5th ed.). Oxford: Oxford University Press.

Cahill Sowle, L. (2005, April 25). Catholicism, death and modern medicine. *America*, *192*, 7–14.

Cataldo, P. J. (1991). Futility and the PVS patient. *Ethics & Medics*, *16*(7), 3–4.

Cataldo, P. J. (1995). The ethical and religious directives for catholic health care services: a commentary. In R. E. Smith (Ed.), *The splendor of truth and health care* (pp. 173–174). Braintree, MA: The Pope John XXIII Medical-Moral Research and Education Center.

Cataldo, P. J. (2004a, Autumn). Pope John Paul II on nutrition and hydration: A change of catholic teaching? *The National Catholic Bioethics Quarterly*, *4*(3), 513–536.

Cataldo, P. J. (2004b, Winter). Queries on nutrition and hydration. *The National Catholic Bioethics Quarterly*, *4*(4), 659–660.

Clarke, S. J. W. N. (2002). *The one and the many: A contemporary thomistic metaphysics*. Notre Dame: University of Notre Dame Press.

Committee for Pro-Life Activities, National Conference of Catholic Bishops. (1992). *Nutrition and hydration: moral and pastoral reflections*. Washington, DC: United States Catholic Conference.

Congregation for the Doctrine of the Faith. (1987, February 22). *Instruction on respect for human life in its origin and on the dignity of procreation*. Available: http://www.vatican. va/roman_curia/congregations/ofaith/documents/rc_con_cfaith_doc_19870222_respect-for-human-life_en.html

Cronin, D. A. (1989). The moral law in regard to the ordinary and extraordinary means of conserving life. In R. E. Smith (Ed) *Conserving Human Life* (pp. 3–145). Braintree, MA: Pope John XXIII Medical-Moral Research and Education Center.

Doerflinger, R. M. (2004, June). John Paul II on the "vegetative state". *Ethics & Medics*, *29*(6), 2–4.

Ford, N. (2004, May 1). *Impacts of papal teaching on vegetative patients in catholic hospitals*. Available: http://www.thetablet.co.uk/articles/2501/.

Garcia, J. L. A. (2008). Understanding the ethics of artificially providing food and water,' In C. Tollefsen (Ed.). *Artificial nutrition and hydration: The new catholic debate* (pp.). New York: Springer, 127–143.

Gilllick, M. (2000). Rethinking the role of tube feeding in patients with advanced dementia. *The New England Journal of Medicine, 342*(3), 206–210.

Grisez, G. (1990). Should nutrition and hydration be provided to permanently comatose and other mentally disabled persons? *Linacre Quarterly, 57*(2), 30–43.

Hamel, R., & Panicola, M. (2004, April 19–26). Must we preserve life? *America, 190*(14), 6–13.

John Paul II, P. (2004, March 20) *Address to the participants in the international congress on "life-sustaining treatments and vegetative state: scientific advances and ethical dilemmas"* [On-line] Available: http://www.vatican.va/holy_father/john_paul_ii/speeches/2004/march/documents/hf_jp-ii_spe_20040320_congress-fiamc_it.html.

John Paul II, P. (2004, November 9). *Letter on the occasion of the 23rd national congress of the italian catholic physicians' association* [On-line] Available: http://www.vatican.va/holy_father/john_paul_ii/letters/2004/documents/hf_jp-ii_let_20041109_medici-cattolici_en.html.

John Paul II, P. (1995). *Veritatis Splendor* [On-line] Available: http://www.vatican.va/holy_father/john_paul_ii/encyclicals/documents/hf_jp-ii_enc_06081993_veritatis-splendor_en.html

Klubertanz, S. J. G. P. (1963). *Introduction to the philosophy of being.* New York: Appleton-Century-Crofts.

Kopfensteiner, T. R. (2000, May–June). Developing Directive 58: A Look at the History of the Directive on Nutrition and Hydration. *Health Progress, 81*(3), 20–23, 27.

May, W. E. (1990). Criteria for withholding or withdrawing treatment. *Linacre Quarterly, 57*(3), 81–90.

May, W. E. (1997). Caring for Persons in the "Persistent Vegetative State". *Anthropotes: Rivista di Studi sulla persona e la famiglia, 13*(2), 317–331.

May, W. E. (1998, December). Tube feeding and the "vegetative" state: part 1. *Ethics & Medics, 23*(12), 1–2.

May, W. E. (1999, January). Tube feeding and the 'vegetative' state: part 2. *Ethics & Medics, 24*(1), 3–4.

May, W. E. (2000). *Catholic bioethics and the gift of human life.* Huntington, IN: Our Sunday Visitor Publishing Division.

McCormick, S. J. R. A. (1978, February). The quality of life, the sanctity of life. *Hastings Center Report, 8*(1), 30–36.

McCormick, S. J. R. A.. (1981a). To save or let die: The dilemma of modern medicine. In *How brave a new world? Dilemmas in bioethics* (pp. 339–351). Washington, DC: Georgetown University Press.

McCormick, S.J. R. A. (1981b). The moral right to privacy. In *How brave a new world? Dilemmas in bioethics* (pp. 362–371). Washington, DC: Georgetown University Press.

McCormick, S. J. R. A. (1992, March 14). "Moral considerations" Ill considered. *America, 166*(9), 210–214.

Meilander, G. (August/September 2004). 'Ethics and Life's Ending: an Exchange,' *First Things, 145*, 31–37.

Murphy, L.M., & Lipman,T.O. (June 9, 2003). 'Percutaneous Endoscopic Gastrostomy Does Not Prolong Survival in Patients with Dementia,' *Archives of Internal Medicine, 163*, 11, 1351–1353.

O'Brien, D., & Slosar J. P. (2004, Winter). The authors reply. *The National Catholic Bioethics Quarterly, 4*(4), 660–662.

O'Brien, D., Slosar, J. P, & Tersigni, A. R. (2004, Autumn) Utilitarian pessimism, human dignity, and the vegetative state: A practical analysis of the papal allocution. *The National Catholic Bioethics Quarterly, 4*(3), 497–512.

O'Rourke, K., O.P., (1988, January–February). Evolution of church teaching on prolonging life. *Health Progress, 69*(1), 28–35.

O'Rourke, K., O.P., (1989, Fall). Should nutrition and hydration be provided to permanently unconscious and other mentally disabled persons? *Issues in Law & Medicine, 5*(2), 181–196.

O'Rourke, K., O.P., (1991, May). Prolonging life: A traditional interpretation. *Linacre Quarterly*, *58*(2), 12–26.

O'Rourke, K., O.P., (1992, July–August). Removing life support: Motivations, obligations. *Health Progress*, *73*(6), 20–27, 38.

O'Rourke, K., O.P., (1999, April). On the care of 'Vegetative' patients. *Ethics & Medics*, *24*(4), 3–4.

O'Rourke, K., O.P., (2002, Winter). Ms. "B" and the vatican. *National Catholic Bioethics Quarterly*, *2*(4), 595–600.

O'Rourke, K., O.P., (2003, Spring). Tube feeding and Ms. "B": father O'Rourke responds. *National Catholic Bioethics Quarterly*, *3*(1), 16–17.

Panicola, M. (2001a, November–December). Catholic teaching on prolonging life: Setting the record straight. *Hastings Center Report*, *31*(6), 14–25.

Panicola, M. (2001b, November–December) Withdrawing nutrition and hydration. *Health Progress*, *82*(6), 28–33.

Pinker, S. (1997, November 2). Why they kill their newborns. *New York Times Magazine*, 52–54.

Robertson, J. (2004, July-August). Extreme prematurity and parental rights after baby doe. *Hastings Center Report*, *37*(4), 32–39.

Shannon, T. A., & Walter, J. J. (1988, December). The PVS patient and the forgoing/withdrawing of medical nutrition and hydration. *TheologicalStudies*, *49*(4), 623–647.

Singer, P. (1993). *Practical ethics*, (2nd ed.). Cambridge: Cambridge Universtiy Press.

Smoker, B. (2003, December). On advocating infant euthanasia (Op-Ed). *Free Inquiry*, *24*, 17–18.

United States Conference of Catholic Bishops. (2001). *Ethical and religious directives for catholic health care services*. Washington, D.C.: United States Conference of Catholic Bishops.

# Part III
# Symposium on the Views
# of Fr. Kevin O'Rourke, O.P.

# Chapter 9
# Reflections on the Papal Allocution Concerning Care For PVS Patients[1]

Kevin O'Rourke, O.P.

## 9.1 Introduction

A recent statement of the Holy Father (John Paul II, 2004) in regard to the care of persons in the persistent vegetative state (PVS) was received with dismay by many people inside and outside the Catholic health care ministry (see O'Brien, 2004; Shannon and Walter, 2004). In sum, the Holy Father stated that artificial nutrition and hydration (ANH) was not medical care, but rather comfort care, and in principle, should be maintained even if there is no hope that the patient will recover from the debilitated condition of PVS. He also maintained that a patient in PVS remains a person in the full sense of the term, something not denied by Catholic theologians, ethicists and care-givers. Finally, the statement indicated that knowingly and willingly removing of AHN from PVS patients is passive euthanasia. Clearly, this was not an infallible or definitive statement of Church teaching, rather it was an authentic or reformable statement (see Code of Canon Law, c. 751, 753; Gaillardetz, 2003, pp. 94–99). The purpose of this essay is to examine the allocution of the Holy Father in light of Church teaching in regard to "reformable statements."

## 9.2 A Consideration of the Norms for Accepting Magisterial Teaching

At one time, the phrase "Roma locuta est, causa finite est," (Rome has spoken, therefore all contrary opinions are over ruled) indicated the proper response for the loyal Catholic theologians. But in 1990, realizing that many times the Holy See

Kevin O'Rourke, O.P.
Neiswanger Institue of Bioethics and Health Policy; Stritch School of Medicine,
Loyola University, Chicago, IL
Email: korourk@lumc.edu

This essay was composed prior to the August 1, 2007 release by the Congregation for the Doctrine of the Faith, "Responses to Certain Questions of the United States Conference of Catholic Bishops Concerning Artificial Nutrition and Hydration." The author has thus not had the opportunity to take this teaching of the Magisterium in account in writing this essay.

has spoken and later reversed its teaching, the Congregation for Doctrine of the Faith, (CDF) issued a statement in regard to the acceptance of Church teaching, often called *Donum Veritatis* (1990).[2] This document explained the responsibilities of theologians and the Magisterium of the Church, showing how the two forms of teaching ministry within the Church should work together. The document outlined four different forms of Magisterial teaching. They are:

(1) "When the Magisterium of the Church makes an infallible pronouncement and solemnly declares that the teaching is found in revelation, the assent called for is of theological faith." Many examples of this infallible form of teaching are found in the Council of Trent, the First Vatican Council, or in the Declarations of the Immaculate Conception and the Assumption of the Blessed Virgin Mary into Heaven, in an "extraordinary" form of teaching. But examples of this form of teaching may also be found in the universal and ordinary teaching authority of the Pope and Bishops, such as the statements concerning abortion and euthanasia in the encyclical *The Gospel of Life* (Ratzinger, 1998).

(2) "When the Magisterium proposes "in a definitive way" truths concerning faith and morals, which even if not divinely revealed are nevertheless strictly and intimately connected with revelation, these must be firmly and accepted and held." The statement of Vatican Council I in regard to papal infallibility and truths of the natural law would fit into this category.

(3) "When the Magisterium, not intending to act "definitively" teaches a doctrine to aid a better understanding of revelation and makes explicit its content, or to recall how some teaching is in conformity with the truths of faith or finally to guard against ideas that are incompatible with these truths, the response called for is that of religious submission of intellect and will (*obsequium intellectus* and *voluntatis*)." As then Father, now Cardinal Dulles, explained, "this third category has long been familiar to Catholics, especially since the popes began to teach regularly through encyclical letters, some two centuries ago. The teachings of Vatican II, which abstained from new doctrinal definitions, fall predominantly within this category" (Dulles, 1991, p. 694). Truths of this nature are often described as "non-infallible" or reformable or authentic teachings. The teaching of the encyclical *On Human Life* of Paul VI in 1968, in regard to moral means of family limitation is of this nature.

(4) "Finally . . . in order to warn against dangerous opinions which could lead to error, the Magisterium can intervene in questions under discussion which involve, in addition to solid principles, certain contingent and conjectural elements. It often becomes possible with the passage of time to distinguish between what is necessary and what is contingent." Cardinal Dulles states that this is a new dimension in Church teaching. Because the recent teaching of the Holy Father in regard to the care of PVS patients contains certain contingent and conjectural elements, the response to this fourth kind of teaching, referred to in the Instruction as prudential teachings, will be our concern.

### 9.2.1 Response to Prudential Teaching

According to the teaching of *Donum Veritatis*, one's first response to this type of teaching is to accept it with submission of intellect and will, (*obsequium intellectus et voluntatis*.) "The willingness to submit to the teaching of the Magisterium on matters per se not irreformable must be the rule." But the teaching in question "might not be free from all deficiencies." It might "raise questions regarding timeliness, the form or even the contents of the magisterial intervention." The Instruction sets forth several prudential norms for re-examining in humility the argumentation that seems to lead to a conclusion contrary to the magisterial teaching. If after a process of this nature, the theologian for reasons intrinsic to the teaching of the document is not able to give intellectual assent to the teaching, "the theologian has the duty to make known to the magisterial authorities the problems raised by the teaching in itself, in the arguments proposed to justify it or even in the manner in which it is presented." It should be emphasized that the reasons prompting the theologian to withhold assent must be "intrinsic to the teaching" to demonstrate that the reasons in opposition to the magisterial teaching must be historically and theologically accurate, not founded merely upon contrary practice or the difficulty of putting the teaching into practice.

In situations of this nature, the theologian should refrain from giving public expression to the difficulties or discrepancies that are found in the teaching and should not turn to the mass media in order to confront the teaching of the Magisterium. "Respect for the truth as well as for the People of God requires this discretion." Private discussion of the teaching, for example with other theologians or even in scholarly journals would not be prohibited. But clearly unsuitable would be any effort to organize vocal opposition or an appeal to rejection of a magisterial teaching through popular opinion. Some might consider this form of response as contrary to the spirit of honesty and openness that should be part of a theologian's character. However, the common good takes precedence over proving the personal opinion of a theologian, no matter how well founded it might be. Thus, there is a possibility for dissent to prudential teachings of the Magisterium described in the Instruction of the CDF. But perhaps dissent is too strong a word. It seems a better word might be "disagreement" or even the phrase, " inability to assent for reasons intrinsic to the teaching." Clearly, to describe the response of a loyal theologian to the teaching of the Church as dissent, might be an exaggeration and also give the impression that the theologian in question is acting in opposition to the Magisterium or has little respect for the role of the Holy Spirit in the life of the Church.

### 9.2.2 A Significant Question

A significant question remains: does the person who is not a theologian but who has some knowledge of the situation to which the teaching applies have the same rights as the theologians described in the Instruction of the CDF? Does a concerned lay person have the same duty as a theologian if he or she perceives from evidence intrinsic to the matter in question that the teaching "might not be free of all deficiencies in regard to timeliness, the form, or even the content of the magisterial intervention."

For example, the teaching in question might be based upon scientific facts or professional practices concerning which the lay person has intimate knowledge. It seems the "ordinary believer" would be able to withhold assent, and to communicate the reasons for this state of mind to the magisterial authority, provided the person in question would follow the same process outlined in the Instruction for the theologian (Gaillardetz, 2003, pp. 121ff). Above all, the inability to assent must be based on well formulated historical and theological reasons and the forum for discussion should not be the mass media. This would preclude basing one's position simply upon the fact that the teaching is difficult to follow, or that many people are engaged in practices opposed to the teaching, as seemed to be the basis for most of the opposition in regard to the teaching of Pope Paul VI contained in the Encyclical *On Human Life*.

### 9.2.3 The Reason for Donum Veritatis

Why was the Instruction *Donum Veritatis* promulgated? In a press conference introducing the document, Cardinal Ratzinger, the Praeses of the CDF, mentioned that several teachings of the Church have been reversed over time: for example, the teaching of freedom of conscience in regard to religion, the separation of Church and State, and many statements of the Pontifical Biblical Commission (Dulles, 1991). Anyone familiar with the papal documents *Mirari Vos* of Gregory XVI and *The Syllabus of Errors* of Pius IX will understand the need for considering this fourth type of Papal teaching (Chadwick, 1998, pp. 23–25, 168–181). Does the recent statement of Pope John Paul II concerning the care of PVS patients fall into the category of statement which might in time be reversed? The main part of this essay will investigate this question; we shall be concerned with an examination of the "contingent and conjectural statements" of the Papal allocution and the suppositions or assumptions upon which they are based. But before proceeding to these considerations, there are two pre-notes which will facilitate our considerations.

### 9.2.4 Two Pre-notes

First of all, we must distinguish clearly between vegetative state (VS) and permanent vegetative state (PVS) because the document under study seems to consider them as one. The Allocution defines vegetative state as a condition in which "the patient shows no evident sign of self-awareness or of awareness of the environment and seems unable to interact with others or to react to specific stimuli." Neurologists would add to this definition the fact that the patient displays sleep-wake cycles; hence, the patients eyes are often open, but unable to track in a meaningful manner. When discussing PVS the Allocution indicates that there is no different diagnosis for it but only "a prognostic judgment that recovery is statistically speaking more difficult." In fact, the transition from VS to PVS is based on more than statistics. It based upon a presumption that the condition of the patient is irreversible, and this presumption is based upon neurological evidence, gained from a lengthy observation of the patient. "Like all medical judgments this presumption is based upon probabilities, not absolutes" (Joint Task Force, 1994).

Secondly, the allocution maintains that decisions to remove life support should not be made on the basis of quality of life "because the intrinsic value and personal dignity of every human being does not change no matter what the circumstances of his or her life." Quality of life is an ambiguous term. Sometimes it is used to signify human dignity, as in the Allocution, but sometimes it is used to signify the circumstances resulting from an illness or pathology. When determining whether or not to utilize or withhold life support, as Pope Pius XII observed, an evaluation of the "circumstances of persons, places, times, and culture" (1958) is necessary before making a decision to withdraw life support. The statement of the Pontifical Council *Cor Unum*, quoted with approval in the Allocution, referred to this analysis of circumstances as judging "the quality of life" (Pontifical Council Cor Unum, 1971). Perhaps when discussing the circumstances which are present in the life of a dying person, we should do away with the term "quality of life," and use the term "quality of function," as suggested by Father Thomas O'Donnell, S.J., a noted medical ethicist, in a private letter many years ago. All persons have the same quality of life because God's love extends to every human person. But all do not have the same quality of function, and it is the quality of function that we evaluate when questions of prolonging life of ourselves or our loved ones must be settled.

## 9.3 Purpose and Content of the Allocution

In the remaining parts of this essay I shall: (i) briefly explain the goals of this Allocution of the Holy Father and express agreement with these goals; (ii) question the assumptions of the Allocution which to my mind are inaccurate and which lead to inaccurate "contingent and conjectural" statements; (iii) present the positive reasons for my personal inability to assent to this papal allocution; and then (iv) discuss some of the implications of the statement if it is taken as a guide for medical, legal and pastoral care. Before beginning this part of the presentation, realize that the document under consideration is the Allocution as issued by the Vatican Press office, not as it has been interpreted by many individuals and agencies.[3]

### 9.3.1 The Goals of the Papal Allocution

Three goals may be discerned from the Papal Allocution:

(1) The Church seeks to counteract the trend in our society and culture toward euthanasia and disrespect for human life. The effort to put people to death in order to end their suffering or to terminate a debilitated existence is demonstrated in law and medicine in the present time. Euthanasia is legal in some countries and states at this time. Pope John Paul II sought to emphasize that life is sacred and to

counteract these vicious trends in the encyclical The Gospel of Life, and carries this message throughout the world on his many journeys.

(2) The Church wishes to speak on behalf of the debilitated and infirm. Above all, the Church seeks to counteract the tendency to have other people decide for the weak and infirm the value of their lives. Fear is expressed that the term "vegetative state" will demean the personal dignity of people in this condition. Thus, the intrinsic value of and personal dignity of debilitated persons is affirmed strongly by John Paul II.

(3) Finally, the Holy Father wishes to stress that no matter how debilitated and bereft of human function, the infirm are still persons, and to be treated as such by medical personnel, families and society.

No one writing from a Catholic perspective, disagrees with the need to work diligently for the attainment of these goals. However, the assumptions upon which a strategy to achieve these goals is based seems subject to question. In the following section I shall consider two of these assumptions and the statements based upon them.

## 9.3.2 Questionable Assumptions that Lead to Questionable Conjectural and Contingent Statements

The Instruction of the CDF states that authentic teachings which contain contingent and conjectural statements may be subject to reversal. According to *Webster's Dictionary* a conjectural statement is one based on incomplete or inconclusive evidence; a contingent statement is likely to be true, but not yet certain; i.e., it is possible but not certain. In so far as papal statements of a prudential nature are concerned the conjectural and contingent statements are based upon assumptions. Thus, an assumption is made that a specific proposition or declaration is true and conclusions are drawn from that assumption; these conclusions are conjectural or contingent, that is, they may or may not be true, depending on the truth of the assumption. In time, an assumption may prove to be untrue, and thus the conjectural or contingent statements which follow from it are also untrue. For example, consider two statements contained in the *Syllabus of Errors* referred to above. In section III of the *Syllabus of Errors*, n. 15, the following statement is condemned: "Every man is free to embrace and profess the religion, which guided by the light of natural reason, he shall consider true." In section VI of the same document, n. 55, the following statement is condemned: "The Church ought to be separate from the State, and the State from the Church." These statements were later reversed by the Second Vatican Council. The Council, in the Decree on Religious Freedom, stated, "that the human person has the right to religious freedom . . . this Council further declares that the right to religious freedom is based on the very dignity of the human person as known through the revealed Word of God and by reason itself" (Flannery, 1980, p. 800, n. 2). The Council also stated, "The political community and the Church are autonomous and independent of each other in their own fields. Both are devoted to the personal vocation of man under different titles" (Flannery, p. 800, n. 76).

The assumptions upon which the statements in the *Syllabus of Errors* were founded are not stated in the original documents. But a knowledge of Church history helps us discern what they were. First of all, the Church had long maintained that it had some kind of power over secular governments. This assumption dates back to the days when Charlemagne was crowned as Holy Roman Emperor by Pope Leo III in 800, and this assumption found full expression in the encyclical *Unum Sanctam* of Boniface VIII in 1302. Moreover, assumptions were present in the Nineteenth Century, when the aforementioned statements were condemned by the Holy See, that if people were allowed freedom of religion or if Church and State were separate entities that people would lose faith in God and the Church, and that the State would persecute and seek to destroy the Church. These assumptions proved untrue and thus the contingent and conjectural statements based upon them were later proven untrue and were reversed by the Church. At the time they were made, these assumptions were questioned by many and the Church suffered embarrassment as a result of the discipline based upon these assumptions (Chadwick, 1998, pp. 168–181).

### 9.3.3  First Assumption of the Allocution

There are assumptions underlying the recent papal Allocution which can be called into question: assumptions which also seem to permeate the thinking of the papal advisors who assisted in the formulation of the allocution.[4] These misleading assumptions lead to contingent and conjectural statements which also may be called into question.

The first assumption that seems to be inconsistent with reality is that there is some hope of benefit from prolonging life for a patient in a permanent vegetative state, even if it is unlikely that the patient will recover. This assumption is held be some theologians and philosophers (May et al. , 1987; Grisez, 1993, pp. 524–526; Boyle, 1995), but is contrary to the opinion of several medical societies that have considered the care of patients this condition (American Academy of Neurology, 1989; American Medical Association, 1992; Multi-Society Task Force, 1994a, 1994b; British Medical Association, 2001), to many theologians and ethicists with clinical experience (O'Rourke, 1989; Broduer, 1990; Paris, 1998; Hamel and Panicola, 2004), and to some members of the hierarchy who have offered guidance to families in specific cases (Gelineau, 1998; Kelly, 1998; Illinois Bishops', 2001). The main support for the opinion that life in PVS is a an "intrinsic good" and a "great benefit" is the conviction of the theologian Germain Grisez and his followers, that human life is an incommensurable good and that those who deny this assertion are professing dualism (O'Rourke, 1989). If human life is an intrinsic good, why does the Church teach that life support may be removed if it imposes an excessive burden? Moreover, as my colleague Benedict Ashley observes:

"the human body is human precisely because it is a body made for and used by intelligence. Why should it be dualism to unify the human body by subordinating the goods of the body to the good of the immaterial and contemplative intelligence" (Ashley, 1994, p. 73).

While it is not a conclusive proof, it is noteworthy that most of the people who maintain that continued existence in a PVS condition is not a "great benefit" have been involved in clinical and pastoral situations. They are not primarily academic persons; they are physicians and ethicists who help families make prudential decisions in difficult circumstances. They realize that when families make decisions to remove AHN from PVS patients that it is not "tantamount to dumping them in the garbage" (Grisez, 1990, p. 40). Finally, Bryan Jennet relates the opinions of several groups of clinical practitioners and lay people in regard to having life prolonged in PVS; which opinions are contrary to the assumption that prolonging the life of PVS patients is a great benefit (Jennet, 2002, pp. 73–86).[5]

This first assumption, that life in PVS is a great benefit even if recovery is highly unlikely leads to a series of contingent and conjectural statements which also can be called into question; statements which seem to remove AHN from the traditional evaluation of hope of benefit because it is presumed that continuation of the vegetative state offers hope of benefit to the patient even though recovery is unlikely. Thus, at best the following statements seem out of touch with reality:

(1) "The evaluation of probabilities, founded on waning hope for recovery when the vegetative state is prolonged beyond a year, cannot ethically justify the cessation or interruption of minimal care for the patient, including nutrition and hydration" (no. IV). In most cases of PVS, moral certitude that the patient will not recover is possible (Multi-Society Task Force, 1994b). This seems to indicate that there is no hope of benefit to the patient if life support is prolonged by means of AHN.

(2) "Death by starvation or dehydration is in fact the only possible outcome as a result of their withdrawal. In this sense it ends up becoming, if done knowingly and willingly, true and proper euthanasia by omission" (no. IV). The disturbing implication of this statement is that it gives the impression that the moral object of a human act is determined by the physical result of the action. This of course is contrary to the teaching of the Church in the encyclical *The Splendor of Truth.* (*Veritatis Splendor*) (John Paul II, 1993, n. 78). The same physical act may have two distinct moral evaluations; e.g. sexual intercourse may be an act of marital love or an act of adultery. The possibility that AHN might ever be withheld or withdrawn is excluded, if the statement in the allocution is taken literally. In this regard, recall the words of the Document issued by the Pro-life Committee of Bishops in the United States, a document in accord with the basic concepts of the Papal Allocution:

"We should not assume that all or most decisions to withhold or remove life support are attempts to cause death. Sometimes other causes may be at work, for example, the patient is imminently dying, whether a feeding tube is placed or not... at other times, although the shortening of the patient's life is one foreseeable result of an omission, the real purpose of the omission was to relieve the patient or the patient's family" (Committee on Pro-Life Activities, 1992, pp. 705ff.).

(3)"Water and food, even when provided by artificial means always represents a natural means of preserving life, not a medical act. Its use furthermore should be considered in principle ordinary and proportionate and as such morally obligatory insofar as and until it is seen to have attained its proper finality which in the present case consists in providing nourishment to the patient and alleviation of his suffering" (no. IV). Even though the papal allocation maintains, in the face of medical and legal opinion to the contrary (Jennet, 2002, pp. 108ff.), that AHN is not medical care, insofar as it "preserves life" it must be morally evaluated by the traditional criteria: hope of benefit and degree of burden (O'Rourke, 1989, p. 194). Moreover, in so far as "finality" is concerned, as we shall see when considering the second assumption, competent medical opinion holds that people in PVS do not experience pain or suffering.

## 9.3.4 The Second Assumption

The second assumption is even more disturbing than the first. It might be phrased in the following manner: "The medical facts and findings of several professional societies, study groups, research papers, court findings and decisions, are not to be considered as valid scientific evidence." This attitude is disturbing because the Holy See usually encourages and values scientific research and seeks to refer to it when issuing instructions or allocutions. The accompanying statement of FIAMC, the World Federation of Catholic Medical Associations, offers inadequate scientific proof for the medical assertions of the Allocution. The above assumption leads to the following statements which are contrary to the finding of several different medical research groups and publications:

(1)"There are a high number of diagnostic errors reported in the literature" (no. II). There are no citations given to "the literature" in question There is no doubt that mistakes of diagnosis are possible, but not if diagnoses are made by board-certified neurologists following the guidelines developed by research groups (Jennet, 2002, chap. 2) People frequently recover from coma and occasionally from VS, but not from PVS which has been properly diagnosed (Levin et al., 1991). As mentioned in the first pre-note, the conditions of coma, VS, and PVS, should not be confused.
(2)"Moreover, not a few of these persons, with appropriate treatment and with specific rehabilitation programs have been able to emerge from the vegetative state . . . We must neither forget nor underestimate that there are well documented cases of recovery even after many years" (no. II). The supposition that recovery from a prolonged vegetative condition or from PVS is likely is also inferred in other parts of the allocution. But on the contrary research publications offer little hope of recovery for PVS patients (Jennet, 2002, chap. 5).
(3)"Moreover, it is not possible to rule out *a priori* that the withdrawal of nutrition and hydration, as reported by authoritative studies is the source of considerable suffering for the sick person, even if we can see only the reaction of the

autonomic nervous system or of gestures." Once again, "the authoritative studies" are not cited in the FIAMC statement. Several contemporary studies maintain that removing AHN from patients in PVS or prolonged coma does not cause pain. In the words of one significant study, "The perception of pain and suffering are conscious experiences: unconsciousness by definition precludes these experiences (Jennet, 2002, pp. 15, 17–18; Multi-Society Task Force, 1994b, p. 1579). With this in mind, describing the removal of AHN as "starving the patient" is a clear misconception.

## 9.4  Positive Reasons for Withholding Assent from the Allocution

The positive reasons for disagreement with the teaching contained in the allocution are founded upon a Thomistic anthropology of the human person. Briefly the goal or purpose of human life is friendship with God; i.e., charity (*Catechism of the Catholic Church*, 1997, n. 1; Aquinas, 1966, *ST* I–II, on charity). In order to strive for this goal, we must perform human acts. St Thomas distinguishes between human acts (*actus humanus*) and acts of man, (*actus hominis*) (1966, *ST* I–II, q. 1, a. 1). Human acts are acts of the intellect and will; acts of man are bodily acts not commanded by the intellect and will, for example, the physiological acts of the body which are not subject to rational activity, such as circulation of blood and digestion. If a person does not have the ability nor the potency to perform human acts now or in the future, then that person can no longer strive for the purpose of human life and it does not benefit the person in this condition to have life prolonged. As Pope John Paul states in the Allocution, "The loving gaze of the Father continues to fall upon them as sons and daughters" but this does not imply that persons in this condition are able to fulfill their part in the reciprocal relationship of friendship, i.e., they are unable to strive for the purpose of life. Therefore, it seems that there is no moral obligation to prolong the life of persons in vegetative state from which they most likely will not recover. Benedict Ashley and I describe the ability to perform a human act as the capacity now, or in the future, to perform acts of cognitive-affective function. If it is morally certain[6] that persons cannot and will not perform acts of this nature now or in the future, then the moral imperative to prolong their lives no longer is present. Hence, it is not "a great benefit" for the patient, for the family nor for society, to prolong their lives. Moreover, health care seeks to help people strive for the purpose of life, not merely to function at the biological level (Pellegrino & Thomasma, 1988, p. 80). Though the sanctity of human life must be affirmed, the fact that death is the gateway to eternal life is often forgotten in contemporary times.

The truth of the foregoing explanation is confirmed by the care given to anencephalic infants in Catholic hospitals. Their lives could be prolonged for a few years perhaps, but no attempt is made to do this. Rather, only comfort care is given because these infants do not have the capacity now or in the future to perform acts of cognitive-affective function. Tube feeding is not part of the comfort care protocol for anencephalic infants. This manner of treatment was approved by a recent statement of the Committee on Doctrine of the United States Conference of Bishops.

Finally, this opinion is based upon the firm conviction that human life is not an absolute good and that there is life after death, when as the liturgy of the Mass for the Dead explains: "Life is changed, not ended." Thus allowing a person to die when continuing efforts to prolong life offers no hope of benefit or imposes an excessive burden is simply surrendering to God's Providence; it is not an act of abandonment.

## 9.5 Implications of the Allocution

This section will merely mention some of the difficulties to which this statement gives rise in order to show the ambiguities of the assumptions and the statements based upon them.

(1) Advance Directives enable people to express their wishes for life support if they are unable to speak for themselves as death approaches. Are these legal documents, which have been approved by many Catholic state conferences in the United States, no longer morally acceptable? Would it be passive euthanasia to be the executor of one of these documents if the person writing it called for removal of AHN when recovery of consciousness seems unlikely?

(2) The allocution seems to imply that financial considerations are not a factor in making prudential decisions about prolonging life. "First of all, no evaluation of costs can outweigh the value of the fundamental good which we are trying to protect, that of human life." And again the questionable statement: "The care of these patients is not in general particularly costly" (see the allocution no. V). Is this in accord with the tradition of the Church in regard to caring for persons with fatal pathologies?

(3) Will Catholic hospitals be required to insure that all patients, families and physicians have AHN utilized for all patients in vegetative state or PVS, even if the people in question are opposed to this form of life support?

## 9.6 Conclusion

A fair question would be: what strategy would be useful to attain the goals mentioned earlier in this paper? The following norms would be significant. First, killing of patients, even in order to alleviate suffering, should be denounced. Secondly, it seems reliance on the traditional and venerable norms for deciding whether or not to use life support, "hope of benefit" and "degree of burden," should be stressed. Third, guidelines for making decisions concerning hope of benefit and excessive burden should be offered but it should be made clear that these decisions are the responsibility of patients and their proxies, designated either by legal document or custom, and that prudential decisions may differ one person to another. Among these guidelines should be the statement that in itself prolonging life for patients in PVS or in a state of prolonged coma is not *ipso facto* a "great good" for the patient.

It seems that the present Directive 58 of the *Ethical and Religious Directives* (ERD) of the bishops of the United States concerning this type of decision is adequate, but it could be enhanced by making it more in accord with the terminology of directives 56 and 57. Thus, it seems the Directive should read: "There should be a presumption in favor of providing nutrition and hydration to all patients, including patients who require medically assisted nutrition and hydration. Specific decisions concerning withholding or withdrawing nutrition and hydration should be made in accord with Directives 56 and 57."

## Notes

[1] This essay originally appeared in *Christian Bioethics* Vol. 12, no. 1, April 2006, and is reprinted with permission.

[2] 'Instruction on the Ecclesial Vocation of the Theologian.' Unless otherwise indicated, all quotations in the text are from this document.

[3] For various interpretations, see, for example, 'Statement of the National Catholic Bioethics Committee,' April 23, 2004; 'Feeding Debate,' *Catholic News Service*, April 7, 2004; John Teavis interviews Bishop Sgreccia and Fathers Mauritio Faggioni, and Brian Johnstone; Richard Doerflinger, *America*, May 3, 2004; Statement of Rev. Dr. Norman Ford, SDB, Chisholm Centre for Health Care Ethics, East Melbourne, Australia, Nicolas Tonti-Phillipini, Canadian Catholic Bioethics Conference 6, 2004; Cataldo (2004, pp. 513–537) and Shannon and Walter (2004, pp. 18–20).

[4] See, for example, Sgreccia Bishop [2004] and Nancy O'Brien [2004].

[5] Other research studies could be cited, but this volume was published recently and contains references to all significant prior studies.

[6] Moral certainty is not equated with physical certainty. Rather, it is the certainty in human affairs from what happens "most of the time" (*ut in pluribus*). CF. *Summa Theologica*, I–II, q. 96, a. 1, ad 3.

## References

American Academy of Neurology. (1989). Position statement on the management and care of the persistent vegetative state patient. *Neurology, 39*, 125–126.

American Medical Association, Council of Ethical and Judicial Affairs. (1992). Decisions near the end of life. *JAMA, 267*, 2229–2233.

Aquinas, T. (1966). *Summa Theologica*, In A. Ross & P.G. Walsh (Eds.), Blackfriars edition. New York: McGraw-Hill.

Ashley, B. M. (1994). What is the end of the human person? The vision of god and integral human fulfillment. In L. Gormally (Ed.), *Moral truth and moral traditions*. Blackrock, IR: Four Courts Press.

Boyle, J. (1995). A case for sometimes tube feeding patients in persistent vegetative state. In J. Keown (Ed.), *Euthanasia Examined* (pp. 189–199). New York: Cambridge University Press.

British Medical Association. (2001). *Withholding and withdrawing life-prolonging medical treatment* (2nd ed.). London: BMJ Books.

Broduer, D. (1990). The ethics of cruzan. *Health Progress, 71*, 42–47.

Cataldo, P. (2004). Pope John Paul II, on nutrition and hydration. *National Catholic Bioethics Quarterly, 4*, 513–537.

*Catechism of the Catholic Church*. (1997). Vatican City: Liberia Editrice Vaticana.

Chadwick, O. (1998). *A history of the popes, 1830–1914*. Oxford: Oxford University Press.

Committee on Pro-Life Activities. (1992). Nutrition and hydration: moral and pastoral reflections. *Origins, 44*, 705–712.

Congregation for the Doctrine of the Faith. (1990). Instruction on the ecclesial vocation of the theologian. *Origins, 20*, 117–126.

Dulles, A. (1991). The magisterium, theology, and dissent. *Origins, 29*, 692–696.

Flannery, A, (Ed.). (1980). *Documents of Vatican II: The conciliar and post-conciliar documents*. Wilmington, DE: Scholarly Resources.

Gaillardetz, R. (2003). *By what authority?* Collegeville, MN: Liturgical Press.

Gelineau, Bishop. (1998). On removing nutrition and water from a comatose woman. *Origins, 17*, 545–547.

Grisez, G. (1990). Should nutrition and hydration be provided to permanently comatose and other mentally disabled patients. *Linacre Quarterly, 57*, 30–38.

Grisez, G. (1993). *Living a christian life*, (vol. 2). Quincy, IL: Franciscan Press.

Hamel, R., & Panicola, M. (2004). Must we preserve life? *America, 190*, 6–13.

Illinois Bishops' Pastoral Letter. (2001). Facing the end of life. *New World Diocesan Paper*, April 15, Chilcago: IL.

Jennet, B. (2002). *The vegetative state, medical facts, ethical and legal dilemmas*. New York: Cambridge University Press.

Joint Task Force. (1994). 'Medical Aspects of the persistent Vegetative State,' *New England Journal of Medicine, 330*, 1499–1508.

John Paul II. (1993). Veritatis Splendor, *Origins, 23*, 297–334.

John Paul II. (2004). Care for patients in permanent vegetative state. *Origins, 33*, 737–739.

Kelly Bishop. (1998). Hugh finn case. Quoted in Paris, J. J. (1998). Hugh Finn's right to Die. *America, 13*, 13–15.

Levin, H. S., Saydari, C., Eisenberg, H. M., Foulkes, M., Marshallm L. F., Ratt, R. M. et al. (1991). *Archives of Neurology, 48*, 580–585.

May, W., Barry, R., Griese, O., Grisez, G., Johnstone, B., Marzen, T. J. et al. (1987). Feeding and hydrating the permanently unconscious and other vulnerable persons. *Issues in Law and Medicine, 33*, 203–217.

Multi-Society Task Force. (1994a). Medical aspects of the persistent vegetative state, part I. *New England Journal of Medicine, 330*, 1499–1508.

Multi-Society Task Force. (1994b). Medical aspects of the persistent vegetative state, part II. *New England Journal of Medicine, 330*, 1572–1579.

O'Brien, N. (April 8, 2004). Some stunned, others affirmed by papal comments on feeding tubes. *Catholic News Service*.

O'Rourke, K. D. (1989). Should nutrition and hydration be provided to permanently unconscious persons? *Issues in Law and Medicine, 5*, 181–196.

Paris, J. J. (1998). Hugh finn's right to die. *America, 13*, 13–15.

Pellegrino, E. D. & Thomasma, D. C. (1988). *For the patient's good: The restoration of beneficience in health care*. New York: Oxford University Press.

Pius XII. (1958). The prolongation of life. *The Pope Speaks, 4*, 395–398.

Pontifical Council Cor Unum. (1971). *Questions of ethics regarding the fatally Ill and the dying*. Vatican City: Vatican City Press.

Ratzinger, C. J. (1998). Commentary on tuendam fidem. *Origins, 28*, 117–119.

Sgreccia Bishop. (2004). Preceding the papal allocution. *Catholic News Service*, March 17.

Shannon, T., & Walter, J. (2004). Implications of the Papal allocution on feeding tubes. *Hastings Center Report, 34*, 18–20.

# Chapter 10
# The Papal Allocution Concerning Care for PVS Patients: A Reply to Fr. O'Rourke

Patrick Lee

In the Allocution of March 20, 2004, Pope John Paul II taught that ANH is to be placed in a fundamentally different category than, say, use of a ventilator or emergency CPR—which are often (though not always) extraordinary and disproportionate care. Rather, ANH is to be considered, in principle, ordinary and proportionate care.[1] Before the Pope's statement, theologians disagreed among themselves on that question (whether ANH is in principle ordinary means or not), and even some bishops and groups of bishops disagreed with each other on the issue. However, Catholics believe that the teachings of the pope are authoritative, even if they are not exercises of extraordinary magisterium and in principle reformable or reversible.

Because the Pope did not teach with his full authority, Catholics owe his statement *religious assent*, as distinct from the assent of faith. This implies that, as the Congregation of the Doctrine of the Faith's document, *Donum Veritatis* points out, such a teaching is to be presumed true, but someone may perceive evidence which makes him unable to assent. Of course, such instances will be infrequent (it is difficult to square regularly disagreeing with the pope, if one believes he is the successor of St. Peter). Still, such a disagreement can occur.

Fr. O'Rourke finds himself in that situation. O'Rourke believes that prolonging the life of a person in PVS does not benefit him and so withholding or withdrawing ANH is often morally correct. O'Rourke's article is addressed to a scholarly audience, and is not part of any public relations campaign to pressure the Pope or to substitute his (O'Rourke's) judgment for the Pope's authority. O'Rourke quotes from *Donum Veritatis,* which says that the teachings of a pope may be based on "certain contingent and conjectural elements" which may turn out to be factually erroneous. Someone who is in a position to see the errors in the assumptions behind the papal teaching, and who perceives evidence for the contradictory of the teaching, may then licitly disagree, and voice his reasons to the magisterial authority.

Patrick Lee
Bioethics Institute, Franciscan University of Steubenville (Director)
Email: plee@comcast.net

C. Tollefsen (ed.), *Artificial Nutrition and Hydration:*
*The New Catholic Debate*, 179–187. © Springer 2008

O'Rourke claims that this teaching on ANH for PVS patients is based on mistaken assumptions, and that there is evidence for the contradictory of the Pope's teaching.

I am not convinced by O'Rourke's arguments, still, since he has only raised questions by presenting them (unlike other theologians, some of whom have denounced the Pope in the mainstream media), I do not believe that he is acting contrary to the duties of a Catholic theologian (except insofar as he advances unsound arguments, and inadvertently mischaracterizes the Pope's position).

In my judgment, O'Rourke is mistaken when he thinks he perceives evidence for the contradictory of the Pope's teaching, and what O'Rourke thinks are mistaken assumptions either are not assumptions (but part of the teaching itself), are not mistaken, or are not premises on which the central papal teaching is based. Since the teaching in this allocution is simply an application of a central and very firm teaching of *Evangelium Vitae*, and O'Rourke has failed to give convincing evidence to contradict it, I am convinced this teaching is correct.

I will first examine O'Rourke's arguments for the contradictory of the Pope's teaching, then examine his claims that the teaching is based on false assumptions, and conclude with an argument that supports the papal teaching.

## 10.1 The Failure of O'Rourke's Arguments Against the Pope's Teaching

O'Rourke claims that his own position is based on a Thomistic anthropology. The purpose of life is friendship with God, but (he argues) to strive for this goal one must be capable of performing *human acts*, that is, acts that flow from intellect and will. He then argues: "If a person does not have the ability nor the potency to perform human acts now or in the future, then that person can no longer strive for the purpose of human life and it does not benefit the person in this condition to have life prolonged" (O'Rourke, 2008, p. 178). He concludes that there is no moral obligation to prolong the life of human beings in vegetative state from which they most likely will not recover. O'Rourke does not claim that such human beings are not persons or that they are not worthy of full moral respect. Rather, his claim is that mere biological life is not a benefit for them.

This argument, however, is unsound for several reasons. It can be formalized as follows:

(1) Only those conditions or actions which contribute to the purpose or goal of the human being are worthwhile.
(2) But mere biological life that does not enable one to perform human acts, either now or in the future, does not contribute to the purpose or goal of the human being.
(Conclusion) Mere biological life that does not enable one to perform human acts, now or in the future, is not worthwhile.

Understood correctly, the first premise is true. But what evidence is there for the second? If an entity is a living bodily being, then how could bodily life not be an important constituent of his fulfillment? A human being is a body-soul composite, not just a soul, and so his bodily life and health must be part of the achievement of his purpose.

O'Rourke's argument for the contrary, that is, for his second premise, is that the purpose of human life is friendship or union with God, but we are brought closer to that goal only by human acts or conditions that enable human acts (now or in the future). However, this rationale for the second premise commits at least one of two errors about the purpose of human life. It conceives the goal of human life as something purely spiritual *or* as something only in the future, or as both. But both assumptions are false.

If friendship with God is conceived as purely spiritual, then the attainment of the purpose of life does and will (in its completion) include more than that friendship—it will also include the resurrected body, and fulfillment of every aspect of human nature, including of course the bodily aspect of the human being. On the other hand, if the *total* condition of friendship or union with God, in the sense that the *Catechism of the Catholic Church* says that friendship with God is the purpose of human life, is referred to, then this condition includes complete human fulfillment, including bodily fulfillment—just as, to work to build up any friendship, conceived broadly, is to work for the well-being of both friends.

Also, even now, in the Eucharist, there is a bodily union with Our Lord, and so we should not view union with God, even narrowly conceived as the relation of union itself, as occurring only in the spiritual aspect of us. Hence the completed kingdom of God, heaven, will include both supernatural communion with God and fulfillment of every aspect of human nature (however that is thought to be related to natural fulfillment).[2] Human life and health are parts of the fulfillment of human life. It is simply false to say that a human individual, who is a body-soul composite, is not benefited by bodily life.

O'Rourke could reply that the purpose or goal of human life is not attained until heaven, which is in the future. And in heaven human fulfillment *will* include bodily life and health, but these will be bestowed by God at the resurrection, and so prolonging the life of someone who cannot in other ways move closer to heaven (by human acts) does not benefit that person. However, it is erroneous to conceive of the purpose or goal of life as something only in the future. That is, in any case, an incoherent notion. While a right *orientation to* something completed only in the future may be a necessary condition for the goodness or appropriateness of a present condition, one's life and present perfection cannot be a mere means in relation to something only in the future. What is only future cannot retroactively bestow meaning or perfection on what is present. Moreover, the Church does *not* teach that the purpose or goal of life is attained only in the future. On the contrary, Our Lord taught that the Kingdom of God is at hand, that is, *now,* though it will be *completed* only in the future. In Vatican II, *Gaudium et Spes*, the Church teaches this point explicitly:

> Therefore, while we are warned that it profits a man nothing if he gain the whole world and lose himself, the expectation of a new earth must not weaken but rather stimulate our concern for cultivating this one. For here grows the body of a new human family, a body

which even now is able to give some kind of foreshadowing of the new age (Vatican Council II, 1965, no. 39).

Thus, the kingdom of God is already present in mystery but will be completed with God's re-creative acts at the end of the world. Even now the goods of this world are constituents of God's kingdom. Although God will re-create the bodily and temporal goods in the future, he will also re-create what we ourselves, with God's grace, build up in this world, though he will re-create these goods without their imperfections.[3] Thus, it is false to say that the purpose and goal of human life is only spiritual—it includes bodily fulfillment. And it is false to say that it occurs only in the future— it begins even now in mystery. Thus, conditions and actions are good not just as means toward some spiritual or future reality. The human body and bodily health, as components of human perfection, are in themselves good.[4]

Nor, contrary to O'Rourke's claim, does his argument accurately reflect Thomistic anthropology. Aquinas does say that the "essence of beatitude (complete fulfill-ment)" is verified in the vision of the divine essence.[5] But he adds that involved in this condition is an overflow (*redundantia*) into all the parts of human nature, so that after the resurrection of the body the completion of every aspect of human nature is part of beatitude in the fullest sense—the "well-being" (*bene esse* of beatitude). So, for Aquinas, beatitude in its most perfect condition includes bodily fulfillment. Moreover, Aquinas also insists that a *participation* or *beginning* of beatitude, which he refers to as *imperfect beatitude*, begins in this life. The constituents of imper-fect beatitude are intrinsically good, that is, good as ends not as mere means, and included in imperfect beatitude are life and health.[6]

After presenting his main argument, O'Rourke says the following:

> The truth of the foregoing is confirmed by the care given to anencephalic infants in Catholic hospitals. Their lives could be prolonged for a few years perhaps, but no attempt is made to do this. Rather, only comfort care is given because these infants do not have the capacity now or in the future to perform acts of cognitive-affective function. Tube feeding is not part of the comfort care protocol for anencephalic infants. This manner of treatment was ap-proved by a recent statement of the Committee on Doctrine of the United States Conference of Bishops (O'Rourke, 2008, p. 178).

However, this is simply false on several points. First, of those anencephalic babies who make it to birth (unfortunately most are aborted) almost all die in a very short time—hours, days, at most weeks, due to causes certainly other than dehydration, probably a combination of causes due to their exposed brain and variably malformed brainstem. Thus, it is false to say that they could be kept alive several years if they were given tube feeding. The experience of neurologists at Loma Linda of plac-ing anencephalic infants on ventilators with the intention of prolonging their lives enough to become organ donors, showed that these infants die early even with venti-lator (which, of course, everyone agrees is, for anencephalic babies, extraordinary or disproportionate means).[7] Second, it is standard practice in most hospitals, at least, to provide these infants with gavage (tube) feeding and hydration (if their sucking reflex is not intact) as part of their comfort care.[8]

Third, the United States Bishops in their statement in 1992 on withdrawing ANH say nothing whatsoever about withdrawing tube feeding from anancephalic infants.[9] Moreover, a more recent statement by the U.S. Bishops on anencephalic infants (1996) explicitly *excludes* the early induction of labor of anencephalic babies as a means of avoiding psychological or physical risks to the mother (Committee on Doctrine, 1996, p. 276). This statement, then, clearly implies that the biological life of the anencephalic child is an intrinsic good to be protected, and so this statement is logically incompatible with the position O'Rourke attributes to the Bishops (namely, that nutrition and hydration could be withheld from anencephalic infants even though it could prolong their lives for years).

## 10.2 O'Rourke's Confusions Regarding Alleged Mistaken Assumptions of the Papal Statement

O'Rourke claims that the papal statement is based on three false assumptions. "The first assumption, that seems to be inconsistent with reality is that there is some hope of benefit from prolonging life for a patient in a permanent vegetative state, even if it is unlikely that the patient will recover." O'Rourke quotes medical societies, theologians and ethicists to support his view that this assumption is false. I contended above that his argument for this claim is unsound, being grounded on a dualistic view of the self (the view that the body is outside the self). As to his appeal to authority, medical societies, theologians, and ethicists can be cited on the other side of this question as well, and—to state the obvious—for a Catholic, papal authority will surely tip the scale to one side rather than the other.

However, most important, the proposition that life for human persons in a vegetative state is a true benefit for them is not an *assumption* of the papal teaching. Rather, it is a *part of* that teaching.

O'Rourke then claims that since this alleged assumption is false, then several other false assumptions follow from it. But since the statement (not an assumption) is true, this claim by O'Rourke is groundless (I shall return in a moment to his rejection of the Pope's teaching that death is often intended in the withdrawal of ANH in a moment).

Next, O'Rourke makes a curious claim: "The second assumption is even more disturbing than the first. It might be phrased in the following manner: 'The medical facts and findings of several professional societies, study groups, research papers, court findings and decisions, are not to be considered as valid scientific evidence'" (O'Rourke, 2008, p. 177). This is O'Rourke's phrasing. But nowhere does the Pope assert anything like this statement. The basis for O'Rourke's remarkable claim here seems to be O'Rourke's disagreement with the following statements of the Pope: (a) There are many diagnostic errors reported in the literature; (b) many of persons in a vegetative state, with appropriate treatment and with specific rehabilitation programs, have been able to emerge from the vegetative state ; and (c) "Moreover, it is not possible to rule out *a priori* that the withdrawal of nutrition and hydration, as reported by authoritative studies is the source of considerable suffering for the sick person" (John Paul II, 2004, no. 3). O'Rourke complains that the FIAMC statement accompanying the papal allocution did not cite the references for these statements.

There are two points to be made about this argument of O'Rourke. First, to rely on some scientists rather than others is very different from ignoring the assertions of medical societies, neurologists, etc.[10] But, second, and most important, the Pope's central teaching does not logically depend upon these statements. All of these statements could be false—though I doubt it—and this would not touch the central *ethical* arguments for the Pope's teaching (the proposition that bodily life is an intrinsic good is not a medical proposition but an ethical one, and so it is not up for grabs scientifically, though A through C, which I just listed above, are strictly scientific or medical statements).

## 10.3 The Central Ethical Argument Demonstrating the Pope's Position

Early in his article O'Rourke says that the allocution stated that "knowingly and willingly removing of ANH from PVS patients is passive euthanasia." But the Pope did not use the term the term "*passive* euthanasia." What the Pope said was that withdrawing nutrition and hydration is often "true and proper euthanasia by omission" (John Paul II, 2004, no. 4). He then added the following:

> In this regard, I recall what I wrote in the *Encyclical Evangelium Vitae,* making it clear that 'by euthanasia in the true and proper sense must be understood an action *or omission* which by its very nature and intention brings about death, with the purpose of eliminating all pain'; such an act is always 'a serious violation of the law of God, since it is the deliberate and morally unacceptable killing of a human person' (John Paul II, 2004, no. 4).

The Pope's point is clear. Generally speaking (that is, not referring to an emergency situation where time and resources may be severely limited) the administration of nutrition and hydration by intubation is not terribly expensive or burdensome. Maintenance of the tube-feeding after its insertion does not require hospitalization and the nutrients are relatively inexpensive. What, then, is the purpose of withholding or withdrawing the tube-feeding in normal situations? Evidently, the purpose is to avoid, not the burdens of the tube-feeding, but the burdens (either on others or on the patient himself) of the continued life of the patient. But the only way to avoid *those* burdens is to bring it about that the patient die. So, often, not providing ANH is chosen as a means toward the dying of the patient: this is, as the Pope says, euthanasia by omission.[11]

The act is analogous to a husband withholding needed insulin from his wife: since the administration of the insulin is not very burdensome, the withholding of insulin is probably chosen as a means of avoiding other burdens, ones associated with the continued life of his wife—but those are avoided only if the wife is dead, so the insulin is probably withheld as a means of bringing about death. In both cases death is intended, and the means chosen to bring it about are omissions.

The central difficulty inherent in O'Rourke's position is again highlighted when he says the following:

> Finally, this opinion [O'Rourke's] is based upon the firm conviction that human life is not an absolute good and that there is life after death, when as the liturgy of the Mass for the Dead explains: 'Life is changed, not ended.' Thus allowing a person to die when continuing efforts to prolong life offers no hope of benefit or imposes an excessive burden is simply surrendering to God's Providence; it is not an act of abandonment (O'Rourke, 2008, p. 179).

"Absolute good" is an ambiguous term. Natural Law theory is concerned with many fundamental human goods (See John Paul II, 1993, no. 48). None of them is "absolute" in the sense that it is morally right to destroy other fundamental goods for its sake, or that it is necessary to take all possible measures to promote or protect it. But each is an intrinsic good and it is wrong to choose to destroy one instance of any fundamental good in order to avoid harm to any other good or to promote other goods. This point is one of the main sources of what are often called "absolute moral norms," that is, specific, exceptionless moral norms. It is true that there is life after death, and it also is true that ceasing *truly* futile efforts to prolong life or withholding *excessively burdensome* treatment is not an act of abandonment. But withholding non-burdensome treatment, on the ground that one has judged someone's life not worth living (and so no longer a benefit) is intentional killing, and the fact that there is an after-life can no more justify this killing than it could other types of intentional killing of innocent people.

Finally, if O'Rourke were correct that the life of patients in PVS is not worth preserving since it is not a benefit to them, then, as William E. May has indicated (May, 2008), it is hard to see why it would not follow that we should withhold food and water (while giving them sedatives to relieve their pain) from other severely mentally impaired individuals who will never in this life be able to perform actions flowing from intellect and will, for example, persons with severe Alzheimer's disease. But that would surely be wrong (and I am sure Fr. O'Rourke would agree), and so too is withholding, in normal situations, ANH from persons in PVS.

## Notes

[1] I take it that since the Pope said "in principle" (John Paul II, 2004, no. 4), there could be exceptions, for example, where the feeding tube would not prolong life very long and the insertion of the feeding tube might itself be extremely difficult; an example might be for certain burn victims.

[2] So this point does not presuppose any particular conception of the relationship between grace and nature.

[3] "For after we have obeyed the Lord, and in His Spirit nurtured on earth the values of human dignity, brotherhood and freedom, and indeed all the good fruits of our nature and enterprise, we will find them again, but freed of stain, burnished and transfigured, when Christ hands over to the Father: 'a kingdom eternal and universal, a kingdom of truth and life, of holiness and grace, of justice, love and peace.' On this earth that Kingdom is already present in mystery. When the Lord returns it will be brought into full flower" (Vatican Council II, 1965, no. 39).

[4] After noting that Germain Grisez says that the life of someone in PVS is an intrinsic good incommensurable to other intrinsic goods, O'Rourke asks: "If human life is an incommensurable good, why does the Church teach that life support may be removed if it imposes an excessive burden" (O'Rourke, 2008, p. x)? The answer is: the fact that a good is intrinsic and incommensurable does not imply that one must take all possible measures to pursue or protect it; one may choose not to take certain measures (such as life support) to realize or preserve one good, even if that results in the diminishing or destruction of that good, if those measures would violate responsibilities toward other goods. Hence the morality of sometimes withholding or withdrawing life support follows, not from the idea that life is an intrinsic and incommensurable good, but from the fact that life is not the only one. I think the fundamental goods *are* incommensurable (which does *not* mean they have no *moral* ordering, since the goods of religion and moral reasonableness should organize one's seeking other goods, only that one fundamental good considered by itself, prior to moral norms, is not more excellent than another, since the fundamental goods are not homogeneous). But the argument I am presenting does not depend on that point, only on the position that it is wrong to choose to destroy, damage, or impede one instance of a fundamental good for the sake of avoiding harm to another, or for the sake of promoting another. See John Paul II, (1993, nos. 48, 52, 67).

[5] The *"consistit in"* in the Latin *"Beatitudo consistit in contemplatione divinae essentiae"* (*Summa Theologiae*, Pt. I–II, q. 3, a. 8) should not be translated as "consists in" but etymologically comes from "stands together in," and so means, *is realized in*, or *is verified in*, rather than *is*.

[6] Compare St. Thomas, *Summa Theologiae*, Pt. I–II, q. 3, a. 2; q. 3, a. 3, ad 2; q. 4, a. 5–6; q. 5, a. 3.

[7] I owe this information to the pediatric neurologist, D. Alan Shewmon.

[8] This fact was confirmed to me by nurses and pediatric neurologists. In a standard text on nursing care of children, Lucille Whaley and Donna Wong said: "Traditionally these infants have been provided comfort measures, but with no effort at resuscitation. Directives to withdraw all support (e.g. feeding) were met with universal resistance by nursery personnel, whose primary function is nurturance" (Whaley and Wong, 1991, p. 471).

[9] Rather, the full paragraph from which O'Rourke quotes is as follows: "Second, we should not assume that all or most decisions to withhold or withdraw medically assisted nutrition and hydration are attempts to cause death. To be sure, any patient will die if all nutrition and hydration are withheld. But sometimes other causes are at work—for example, the patient may be imminently dying, whether feeding takes place or not, from an already existing terminal condition. At other times, although the shortening of the patient's life is one foreseeable result of an omission, the real *purpose* of the omission was to relieve the patient of a particular procedure that was of limited usefulness to the patient or unreasonably burdensome for the patient and the patient's family or caregivers. This kind of decision should not be equated with a decision to kill or with suicide" (Committee on Pro-Life Activities, 1992, no. 1).
Their point is that in some cases of withdrawing ANH the purpose is to avoid the burden of the feeding or insertion of the feeding, where the patient will die very shortly any way, and so the nutrition and hydration might be of insignificant benefit, and so in such cases the withholding or withdrawing ANH does not involve intentional killing. In the very next paragraph the Bishops contrast that case with the wholly different type of case where the purpose is to avoid the burden of the continued life of the patient. In that situation, they say, the death *is* intended and it is immoral: "The harsh reality is that some who propose withdrawal of nutrition and hydration from certain patients do directly *intend* to bring about a patient's death, and would even prefer a change in the law to allow for what they see as more 'quick and painless' means to cause death. In other words, nutrition and hydration (whether orally administered or medically assisted) are sometimes withdrawn not because a patient is dying, but precisely because a patient is not dying (or not dying quickly) and someone believes it would be better if he or she did, generally because the patient is perceived as having an unacceptably low 'quality of life' or as imposing burdens on others." *Ibid.*

[10] D. Alan Shewmon has shown that some individuals with complete lack of cortical function nevertheless exhibit behavior demonstrating that they perceive and feel. Hence lack of cortical functioning

(often taken diagnostically as a sufficient sign of vegetative state) does not necessarily lead to lack of awareness or consciousness (nor to lack of ability to feel pain). He also discusses several significant definitional and methodological problems with the Report of the Multi-Society Task Force on PVS, see Shewmon, (2004). Nevertheless, the conclusion of that task force was as follows: "We believe there are sufficient data on the prognosis for neurologic recovery to allow us to distinguish between persistent and permanent vegetative states. These data, in conjunction with other relevant factors in an individual patient, can be used by a physician to determine when the persistent vegetative state becomes permanent—that is, when a physician can tell the patient's family or surrogate with a high degree of medical certainty that there is no further hope for recovery of consciousness *or that, if consciousness were recovered, the patient would be left severely disabled.*" (emphasis added) Although Shewmon casts doubt on the certainty of diagnoses of "permanent vegetative state," this statement does not assert as much confidence in reaching certainty about the permanence of the vegetative state as first appears, given the last, very significant clause (the patient will be permanently unconscious *or* severely disabled). Moreover, the task force even adds the following: "An accurate diagnosis is critical. Errors in diagnosis have occurred because of confusion about the terminology used to describe patients in this condition, the inexperience of the examiner, or an insufficient period of observation." *Ibid.* Thus, the scientific record on this issue is not as clear-cut as O'Rourke assumes.

[11] Hence it is inaccurate to suggest, as O'Rourke does, that the Pope's analysis "gives the impression that the moral object of a human act is determined by the physical result of the action" (O'Rourke, 2008, p. x). Rather, the moral object is determined by what is the proximate object of the will. If the benefit of the omission is only the avoiding of the burdens of the prolonged life, as opposed to the burdens of the treatment, then the omission is chosen as a means of bringing about death, since death is necessary to avoid the burdens of the prolonged life.

It should also be noted that someone may have an emotional reaction to ANH and choose to withhold or withdraw it on the basis of that emotional repugnance, but a morally good choice is based on intelligible goods and bads, not purely on emotional repugnance.

# References

Committee on Doctrine. (1996). Moral principles concerning infants with anencephaly. *Origins, 26,* 276.

Committee on Pro-Life Activities. (1992). Nutrition and hydration: Moral and pastoral reflections. *Origins, 44,* 705–712.

John Paul II. (1993). Veritatis splendor. *Origins, 23,* 297–334.

John Paul II. (2004). Care for patients in a "Permanent" vegetative state. *Origins 33,* 737, 739–740.

May, W. E. (2008). Caring for persons in the "persistent vegetative state" and Pope John Paul II's March 20 2004 address, "on life sustaining treatments and the vegetative state. In C. Tollefsen (Ed.), *Artificial nutrition and hydration: the new catholic debate* (pp. 63–77). Dordrecht: Springer.

O'Rourke, K. O.P. (2008). Reflections on the papal allocution concerning care for persistent vegetative state patients. In C. Tollefsen (Ed.), *Artificial nutrition and hydration: The new Catholic debate* (pp. 169–181) Dordrecht: Springer.

Shewmon, D. A. (2004). The ABC's of PVS, problems of definition. In C. Machado, & D. A. Shewmon (Eds.), *Brain death and disorders of consciousness* (pp. 215–238) New York: Kluwer.

Vatican Council II (1965). *Gaudium et spes.* Vatican City: Libreria Editrice Vaticana.

Whaley, L. F. & Wong, D. L. (1991). *Nursing Care of Infants and Children* (4th ed.). St. Louis: Mosby.

# Chapter 11
# Response to Patrick Lee

Kevin O'Rourke, O.P.

Patrick Lee has taken the time and energy to respond to my essay concerning the Papal Allocution of March 20, 2004, "On the Care of Persons in a Vegetative State." With consideration and acuity, he states correctly that my concerns are expressed with a view to following the norms set forth for response to papal teaching, as outlined in the document *Donum Sapientiae*, of the Congregation for Doctrine and Faith in 1990. For this reason, my essay, written two years ago and sent to the proper Vatican dicastery, appears for the first time in a volume considering this topic.

Moreover, in expressing his specific disagreements with the reasoning set forth in my essay, Lee evidences a grasp of my thinking and a calm response to it. He does not ask me to read once again the statement of Pope John Paul II, as did one person responding to my thoughts on this matter. Perhaps he realizes that I have read the Papal Allocution at least one hundred times, and in five different languages. In an effort to respond to Dr. Lee's thoughts, I shall be as brief as possible, hoping to maintain the calm and respectful manner which he has initiated.

Certainly, there are many particular statements in regard to care for the dying upon which Dr. Lee and I would agree. We agree that upon occasion, life support in the form of artificial nutrition and hydration (ANH), may be withdrawn from patients in a permanent vegetative state (PVS). He states correctly that because the Papal Allocution uses the term "in principle," (or as the French version has it, "as a general rule") that there are exceptions to the proposition that ANH is ordinary and proportionate care.

Our disagreement concerns the conditions that justify these exceptions. Following the traditional teaching of the Church, stated for example in the ERD, D. 56 and 57, many persons with the same interpretation as I, maintain that life support may be withheld or withdrawn if a patient (or the proxy) determines that it offers no hope of benefit or imposes an excessive burden. Moreover in accord with D. 58, we agree that there should be a presumption to use ANH for patients who need it, but that the presumption ceases when the burden for the patient outweighs the

Kevin O'Rourke, O.P.
Neiswanger Institue of Bioethics and Health Policy; Stritch School of Medicine,
Loyola University, Chicago, IL
Email: korourk@lumc.edu

C. Tollefsen (ed.), *Artificial Nutrition and Hydration:*
*The New Catholic Debate*, 189–191. © Springer 2008

benefit. According to our position, maintaining the life of a person in a permanent vegetative condition with ANH, when there is no hope of recovery, is not of benefit for the patient.

Dr. Lee, following the thought of Germain Grisez, would maintain that prolonging the life of a person in this condition, is of *great benefit* to the patient. This disagreement is the result of two different theological views of the goods of human life. Personally, I think the resolution of this disagreement depends upon an interpretation of the Thomistic concept of the purpose of human life, as discussed in my essay

Moreover, Dr. Lee, is in agreement with some statements made by physicians at the conference in Rome which resulted in the papal allocution under discussion. Hence, he is reluctant to admit that the diagnosis of permanent vegetative state can be made with moral certainty, and thus he would agree that people often recover from this condition. On the other hand, basing my conviction on the findings of board certified neurologists, I believe that the condition of permanent vegetative state can be diagnosed with moral certitude. Note for example the case of Terri Schiavo; the autopsy more than confirmed the diagnosis made before her death by board certified neurologists. Clearly, my position admits the need for caution in making a PVS diagnosis, and allows that a Minimal Conscious State (MCS) might develop. But when there is moral certitude from clinical evidence that the patient is in a PVS condition, it is safe to say that ANH is no longer "in principle" an ordinary and proportionate care. Resolution of this disagreement is more within the realm of medical research than within the realm of theological investigation.

In the course of his considerations, Dr. Lee puts forth two arguments to which I would respond briefly in hopes that these arguments can be put in mothballs. First of all, following the thinking of Grisez and William E. May, he states that it is *dualism* to maintain that when there is moral certitude that a person will not recover or develop the potency to perform human acts, it is no longer necessary to seek to prolong that person's life. This argument is inaccurate and fallacious. If anyone can be accused of dualism, it would be Grisez and followers because they make the "mere biological function of the body," (Lee's terminology) an absolute value, even when it is morally certain the person in this condition will not recover rational activity. All holding a position similar to mine would agree that biological function of the human body is vital for pursuing the purpose of life. But preserving "mere biological function" does not help the patient pursue this purpose. In response to the charge of dualism, I will repeat the thought of my colleague Benedict Ashley;

> the human body is human precisely because it is a body made for and used by intelligence. Why should it be dualism to unify the human body by subordinating the goods of the body to the good of immaterial and contemplative intelligence?" (Ashley, 1994, p. 73).

Secondly, Dr. Lee states that " the administration of nutrition and hydration is not expensive or terribly burdensome. Maintenance of the tube-feeding after its insertion does not require hospitalization and the nutrients are relatively inexpensive" (Lee, 2008, p. 188). The usual care for PVS patients is in hospital or a long term care facility, and it is very expensive. If it is so easy to care for permanently unconscious patients at home, why don't more families do it? Even if a family occasionally seeks to care for a PVS patient at home, it requires constant changing and turning the patient. If people who hold that caring for persons in PVS is not "terribly burdensome" would do some of the nursing required for these patients, then their attitudes would be more credible.

# References

Ashley, B. M. (1994). What is the end of the human person? The vision of god and integral human fulfillment. In L. Gormally (Ed.), *Moral truth and moral traditions*. Blackrock, IR: Four Courts Press.
Lee, P. (2008). The papal allocution concerning care for PVS patients, a reply to Fr. O'Rourke. In C.Tollefsen (Ed.), *Artificial nutrition and hydration: The new Catholic debate* (pp. 183–191). Dordrecht: Springer.

# Chapter 12
# The Morality of Tube Feeding PVS Patients: A Critique of the View of Kevin O'Rourke, O.P.

**Mark S. Latkovic, S.T.D.**

This article will examine and critique the influential position of the eminent moral theologian Kevin O'Rourke, O.P. on the morality of tube feeding patients said to be in a "permanent vegetative state" (PVS).[1] My criticism will primarily focus on a representative sample of Fr. O'Rourke's published articles expressing his views over the last twenty years. These writings – sometimes co-authored with his distinguished colleagues Benedict Ashley, O.P. and Jean DeBlois – and the many responses to them show that O'Rourke has consistently maintained that caregivers are not morally obliged to feed PVS patients – whether orally or by tube. I will also draw on material from an April 2005 debate in Detroit that I moderated between O'Rourke and William E. May,[2] as well as recent news reports of O'Rourke's negative reaction to both Pope John Paul II's March 20, 2004 address on the question of tube feeding PVS patients[3] and to the case for tube feeding the severely brain-damaged Florida woman Terri Schiavo (whose own alleged PVS was highly disputed). I have chosen to focus primarily on O'Rourke because of his very vocal and public criticism of the Pope's address. In light of the late Holy Father's authoritative intervention on the issue and the March 2005 starvation death of Mrs. Schiavo in Florida after her feeding tube was removed, O'Rourke's position is deserving of renewed moral scrutiny.[4]

First, I will summarize O'Rourke's position on the question. This will include analyzing his basic anthropological and moral presuppositions. Second, I will criticize O'Rourke's position by briefly comparing and contrasting it with the Pope's address and with other authors who have made the case for tube feeding PVS patients in most cases. Thus, I hope to both substantively contribute to the moral discussion and to be among the first to offer a criticism of O'Rourke's position that takes into account the Pope's address and his reaction to it. Needless to say, this article is not a comprehensive treatment of the issue, but rather a more limited critiquing of one very influential theological view supporting the denial of tube feeding to PVS patients. Thus, for example, it does not intend to offer either a full-blown defense

Mark S. Latkovic, S.T.D.
Sacred Heart Major Seminary, Detroit, MI
Email: latkovic.mark@shms.edu.

C. Tollefsen (ed.), *Artificial Nutrition and Hydration:*
*The New Catholic Debate*, 193–209. © Springer 2008

of tube feeding PVS patients or a defense of the Pope's address. It is, moreover, the view of a moral theologian who claims no particular medical expertise in the matter of tube feeding or care for PVS patients.

## 12.1 O'Rourke on Why Feeding and Hydrating the PVS Patient is Non-obligatory

### 12.1.1 The "Spiritual Purpose of Human Life" Argument

The starting point for understanding O'Rourke's position on tube feeding and the PVS patient is his long-standing interpretation – worked out with Benedict Ashley, O.P. – of Pope Pius' XII's 1957 address on the "spiritual end" of human life and the criteria for withholding or withdrawing life support. This interpretation, which has been the subject of intense debate, involves questions of theological anthropology. Pius XII declared:

> Normally one is held to use only ordinary means [to sustain life] – according to the circum-
> stances of persons, places, times, and cultures – that is to say, means that do not involve any
> grave burdens for oneself or for another. A more strict obligation would be too burdensome
> for most people and would render the attainment of higher, more important goods too diffi-
> cult. *Life, health, all temporal activities are in fact subordinated to spiritual ends* (Pius XII,
> 1957).[5]

Ashley and O'Rourke interpret this statement to mean that prolonging life is a benefit "only when it gives the person opportunity to continue to strive to achieve the
spiritual purpose of life" (Ashley and O'Rourke, 1997, p. 426).[6] In order to strive
for the spiritual end of life, they add, one needs "some degree of cognitive-affective
function." Thus, they conclude, to determine whether a treatment is of benefit to
someone, "mere physical survival" at some minimal level must be assessed in terms
of the "possibility of spiritual advance it offers" (Ibid.).

If it remains possible in some fashion to perform acts of knowing and loving God
and neighbor then the artificial prolongation of life by medical means is justified;
when this becomes impossible "because consciousness and freedom have been irre-
vocably lost by deterioration of the brain," then such means cease "to be of any real
benefit to the subject." This position, they argue, is in keeping with their teleological
understanding of the Christian moral life: "that all of our free decisions must be
measured by our ultimate goal, eternal life with God" (Ibid.).[7]

According to Ashley and O'Rourke, although human life on earth is a "great
good," making possible our pursuit of other goods, it is neither absolute nor ultimate.
Thus, they tend to see human life as "basic" but also "instrumental." They also
write: "Moral theology has always held that 'benefit' for the human being cannot
be measured in merely physical terms, but rather in terms of the goal of the total
person, soul and body. *Physical life* is a great human value, but it is *subordinated to
the eternal destiny of the whole person*." Its existence, therefore, cannot outweigh
"all burdens required to preserve it" (Ashley and O'Rourke, 1997, p. 425).

With regard to medically assisted nutrition and hydration to PVS patients, the authors, in a section summarizing various views, argue – not surprisingly, given their understanding of physical, bodily life: "if cognitive-affective function cannot be restored, the *artificial* nutrition and hydration may be withheld or withdrawn because there is no moral obligation to continue using ineffective medical means" (Ashley and O'Rourke, 1997, p. 426).[8] While they are sensitive to those Church documents that speak of a "presumption in favor of providing hydration and nutrition to all patients," (Ibid., p. 426) because of the abuses that can happen in end-of-life care, nonetheless they say "presumptions yield to facts, and consequently the real issue remains the estimation of benefits and burdens involved" (Ibid.). So, when a PVS or other grave condition has been diagnosed with a high level of certainty as ruling out a patient's ability to perform human acts, "further life support can only be judged 'aggressive' and 'extraordinary,' and thus ceases to be obligatory." "Such care by its very nature," they continue, "is highly burdensome to the caregivers and is an indignity to the patient. 'Normal' or 'comfort care' that avoids any pain or indignity to the dying patient always remains obligatory," but, they conclude, it "need not include the continuation of nutrition and hydration by intubation" (Ibid.).[9]

## 12.1.2 Some Further Clarifying Points

In order to understand O'Rourke's position on the care of PVS patients, it will be helpful to note a number of other clarifying points that apply to the seriously ill and dying, and not just to PVS patients. First, he and Ashley insist that evaluating whether medical means to prolong life are "ordinary" or "extraordinary," "one must consider the condition of the patient as well as all the social and familial circumstances" (Ashley and O'Rourke, 1997, p. 420).[10] Therefore, to speak even of feeding a patient or any other form of care/treatment as "ordinary"/"extraordinary" apart from knowing his or her condition, is impossible.

O'Rourke, in particular, has been quite clear in saying, in many of his writings on the meaning of "ordinary" and "extraordinary," that Pius' XII's statement presents "the spiritual goal of life as the norm for judging whether a grave burden is present." Thus, he understands ordinary measures of life support as those "which are obligatory because they enable a person to strive for the spiritual purpose of life," while extraordinary measures of life support are those "which are optional because they are ineffective or a grave burden in helping a person strive for the spiritual purpose of life" (O'Rourke, 1988, p. 32).

Second, against those who would apply the benefit and burden assessment only to medical treatment (which they say can sometimes be optional) and not to so-called "basic" or "comfort care" (which they say is always required) Ashley and O'Rourke argue that medical treatment "can also be a form of 'normal care,' " while tube feeding "is obviously a form of medical intervention." For them, the real ethical question "is whether a particular form of 'care' is of real benefit, no matter whether it be 'medical' or not" (Ashley, 1997, p. 421).[11]

Third, they explicitly affirm the following:

> No class of persons should be refused therapy. The decision to withhold or remove life
> support should not be made *unless a fatal pathology is present*; that is, an illness or disease
> from which the person will die if therapy is not utilized. Hence, life support should not
> be withheld from the retarded or severely debilitated simply because they are disabled"
> (Ashley and O'Rourke, 2002, p. 190).[12]

The reference to a "fatal pathology" above might seem to indicate that Ashley
and O'Rourke would be willing, at least *in principle*, to tubally feed and hydrate
PVS patients. However, that is not the case. In addition to the "spiritual purpose
of life" argument, O'Rourke is well known for his "fatal pathology" argument (see
O'Rourke, 1986, 321–323, 331).[13] Although this is sometimes identified as a separate justificatory argument for withholding or withdrawing food and water that is artificially provided, (See Berkman, 2004, pp. 89–90)[14] the way Ashley and O'Rourke
employ it in their joint writings, the argument actually forms, I believe, along with
the spiritual end of human life argument, one single rationale for the non-obligatory
nature of tube feeding. That is, PVS patients are unable to pursue the purpose or end
of human life precisely because they do in fact have a fatal pathology, in this case the
inability to chew and swallow. Thus, says O'Rourke, one is morally justified in allowing the "existing fatal pathology . . . to take its natural course" (O'Rourke, 1986,
p. 322) by withdrawing or withholding artificially administered food and water. If a
minimum degree of cognitive-affective function "in an adult cannot be restored or if
an infant will never develop this function, and if a fatal disease is present, the adult
or infant may be allowed to die" (O'Rourke, 1988, p. 33).

   In their co-authored bioethics textbook, Ashley and O'Rourke admit that while ir-
reversibly unconscious persons (e.g., PVS patients) may be "physiologically
'stable' " for an indefinite period of time, "they are in fact 'terminally ill' and really
'dying,' " although not immanently. "In the normal course of nature their pathology
is actually causing their death, although that dying process has been slowed to a
snail's pace by the life-support technology. When these procedures cease, the person
dies directly from the existing pathology and only indirectly from the withdrawal of
life support" (Ashley, 1997, p. 423). This withdrawal – even of artificially supplied
food and water – could be morally justified based on the principle of double effect
(Ibid., p. 425).

## 12.2 Reasons Why O'Rourke's Position is Wrong and the Address of John Paul II is Right

### 12.2.1 Responding to the Arguments of O'Rourke

John Berkman has noted, however, that "fatal pathology" can be understood in two
senses. First, it could mean "fatal if no treatment is given." But then the argument
would establish little if anything. For, as Berkman says, "without someone having at
least a potentially fatal pathology, the conversation concerning the duty to preserve

one's life never arises." Nor does the presence of a fatal pathology provide moral criteria assessing treatment options. Secondly, "fatal pathology" could mean "fatal regardless of the treatment given." If it means the latter, "then this would seem to mean that the patient is imminently dying, or at least terminally ill." These terms, for Berkman, "more unambiguously constitute a prognosis of a particular patient's condition than does 'fatal pathology,' and thus function better as criteria for evaluating the choice to withhold or withdraw" medically assisted nutrition and hydration (Berkman, 2004, p. 90).

Berkman argues that this distinction illumines the differences between the Catholic theological discussion in the 1950's of medically assisted nutrition and hydration for patients in a "terminal coma" – where this condition was at the time considered "imminently dying" – and the debate as it proceeded in the 1980's, where "whether for good or ill, PVS patients could not for the most part be accurately defined as being imminently dying or even terminally ill" (Ibid.). The fact that PVS patients are neither immanently dying nor terminal is a point that William E. May has stressed in many articles criticizing the O'Rourke position (and those similar to it) over the years.[15] Judgment about whether a fatal pathology is present, let it be noted, should be made *before*, not after life support has been utilized.[16] Moreover, as Eugene Diamond has pointed out, contra O'Rourke, "patients in the so-called [PVS] are able to swallow their own secretions in most cases. The nerves and muscles required for mastication and deglutition are characteristically intact" (Diamond, 1999, p. 2). He also notes that a recent study of forty-three PVS patients in London published in the *British Medical Journal* (July 1996) "indicated that some had cognitive impairment but were still able to communicate with therapists and even do simple computations. Over half [of them] regained consciousness eventually." Thus, he significantly concludes, "there is evidence for cognitive and affective function in PVS" (Ibid.).[17]

Even if one were still to claim that PVS patients were suffering from a fatal pathology (understood as, e.g., terminal cancer, the final stages of AIDS, and *not* the PVS itself), I would argue that there are good reasons for regarding medically assisted nutrition and hydration as *ordinary* and *basic care* that, unless the patient is *immanently* dying from the fatal disease, should be given when the patient's digestive tract can assimilate it (thus it is useful in keeping the patient alive) and when it is not overly burdensome to him or her. As John Keown has convincingly argued, with reference to the English PVS patient Tony Bland – an early 1990's case eerily similar to that of Terri Schiavo – inserting "a gastrostomy tube into the stomach requires a minor operation which is clearly a medical procedure. But it is not at all clear that the insertion of a nasogastric tube is a medical intervention. It is, after all, something that even relatives can be shown how to do . . . The question in such a case [as Bland] is surely why *the pouring of food and water down the tube* constitutes medical treatment. What is the pouring of food and water supposed to be treating" (Keown, 2002, p. 219)?[18]

But what of the claim of Ashley and O'Rourke that if a respirator can be withdrawn as a form of extraordinary *medical treatment* (and as morally acceptable in certain cases according to Catholic moral teaching as it was in the Quinlan case), so too can tube feeding?[19] Keown observes, however, that "ventilation replaces the patients capacity to breathe whereas a tube does not replace the capacity to digest and merely delivers food to the stomach . . . A feeding-tube by which liquid is delivered to the patient's stomach is, it could be reasonably argued, no more medical treatment than a catheter by which it is drained from the patient's bladder" (Keown, 2002, p. 220).[20] The larger question, however, that Keown raises is this: Even if one regards tube feeding as medical treatment, why is it useless? Is it because it will not restore PVS patients to some decent level of health, consistent with the purposes of *medicine*? Or is it because the *lives* of these patients are considered useless?[21]

If one applies the theological anthropology of Ashley and O'Rourke as they themselves apply it to the PVS patient, then it seems that the answer to Keown's question would be: the treatment is useless because his or her life is futile, that is, of no benefit, not worth living. This judgment rests, I believe, on a dualistic understanding of the person and can border on a "quality of life" ethic. A number of authors have noted this, among them William E. May, John Berkman, and Peter Ryan, S.J. Berkman summarizes well the criticisms – which I agree with – of the "spiritual purpose of life" argument. Critics of the latter, he notes, claim "that it assumes a dualistic anthropology, requiring persons to disassociate 'themselves' and their spiritual purpose from their character as bodily creatures.[22] Critics further note that humans are not 'in' their bodies, but that their bodies are in some sense constitutive of who they are." Moreover, Berkman continues, the "spiritual purpose of life" argument "typically assumes functional criteria for 'personhood' and thus leads to the exclusion of certain classes of human beings from care typically extended to all persons." Hence not only PVS patients could be denied medically assisted feeding, but also, according to the logic of the argument, "various classes of patients who through genetic disease or other debility are unable to perform human acts." Since tube feeding or other medical treatments will not benefit them (i.e., by enabling them to pursue the spiritual end of life), "there is no purpose to treating them should they develop any kind of life-threatening (but manageable) illness" (Berkman, 2004, pp. 91–92).

Hence, as William E. May has said many times, Ashley's and O'Rourke's reasoning would, as an example, wrongly justify the decision not to stop arterial bleeding in an infant suffering from trisomy 13, who has no cognitive abilities. For simply stopping the infant's bleeding would be *ineffective* in enabling him or her to pursue the spiritual goal of life and therefore it would be *extraordinary* in their analysis, if they are consistent. According to May, however, only treatments "which would *prevent* a person from pursuing the spiritual goal of life, *disabling* him or her, would be truly 'extraordinary' because of the terrible burden it would impose upon them" (May, 2000, p. 256). In their Detroit debate, May pressed O'Rourke on this point and he responded by saying that infants with trisomy 13 are still capable of "affection." This of course is true. But it is a far cry from the role that the ability to perform human acts plays in Ashley's and O'Rourke's theological anthropology. For recall that Ashley and O'Rourke claim that in order to pursue the spiritual goal of life one must be able to engage in *actus humanus*, that is, acts *requiring not merely "cognitive/affective" abilities (which little babies and baboons have) but intelligence and free choice or the use of reason.* I am not saying that O'Rourke would refuse to bandage the wound of the trisomy 13 infant (I am sure that he would, as he said in the Detroit debate). But there is nothing in his interpretation of Pius XII's address that would give him a principled *reason* for doing so.

Ryan argues that although O'Rourke views human life as a "basic good," he "fails to recognize that bodily life is intrinsically valuable; he regards it only as instrumentally valuable and considers it worthwhile only insofar as it enables a person to participate in other goods" (Ryan, 2000),[23] for example, to pursue life's spiritual goal. This leads O'Rourke, I believe, to fall, however reluctantly and even inconsistently in some ways with his own anthropological principles,[24] into "quality of life" judgments which focus not so much on the objective usefulness of the *treatment*, but on the value of the *life* that is to be preserved or not preserved; not on whether the treatment is burdensome, but on whether the person's life is burdensome.[25] Morally sound "quality of life" judgments, on the other hand, are those "which bear on the usefulness or burdensomeness [these being the two main traditional criteria for determining 'ordinary' and 'extraordinary' treatment] of specific kinds of *treatments* for persons in *specific kinds of conditions*." These "are not the same as 'quality of life' judgments asserting that those persons' *lives* are no longer of any value" (May, 2000, p. 258, note omitted).

Ryan also tellingly shows that O'Rourke's "conception of knowing and loving God [i.e., the "spiritual end of life"] is very narrowly drawn." Catholic pastoral practice, he reminds us, "treats those who apparently do not have that capacity or potential as apt recipients of the sacraments. Just as the severely retarded are regularly baptized, so also PVS patients, like patients in a coma and on the verge of death, are apt recipients of Baptism, Confirmation, and the Sacrament of the Sick." This pastoral practice, Ryan continues, "presupposes that the relationship these recipients have with God can be affected by receiving these sacraments. Moreover, to agree that the purpose of life is to know and love God does not entail holding either that fundamental human goods, including the good of bodily life itself, lack intrinsic value or that the purpose of life does not include participating in those goods" (Ryan, 2000).

O'Rourke, I think, would respond to this last point by saying that to consider human life as an intrinsic good does not imply that it must be kept alive "as long as is physically possible." He thinks that some that make this claim (e.g., William E. May) are influenced by Germain Grisez's moral theory, specifically his idea of (eight) incommensurable goods which he says are non-hierarchically ordered and "independent" of each other, and that this departs from the thought of St. Thomas Aquinas, who held that "the goods toward which we have a natural inclination may at times be subordinate to other human goods." Proximate goods, according to Aquinas, says O'Rourke, "are subordinate to our final goal or good . . ." and they "are to be evaluated morally in so far as they are proportionate to striving for the objective ultimate good: knowing and loving God" (O'Rourke, 1988, part 2, p. 3).[26]

Here is not the place to fully engage the contested argument over the question of the hierarchy of goods in Aquinas' natural law theory. Nonetheless, I maintain with Grisez and John Finnis[27] that St. Thomas does *not* arrange the goods to which we have a natural inclination to in a hierarchy and that it can be dangerous to do so.

For example, it can be dangerous when a so-called lesser good of human persons, such as human life, is no longer deemed worthy of protection because it itself can no longer serve as the instrument for realizing the so-called higher goods.[28]

A better moral and anthropological approach would be simply to say, with John Paul II, as he does in *Veritatis splendor* no. 13, when commenting on the story of the encounter between Jesus and the rich young man, that the commandments of which Jesus "reminds the young man are meant to safeguard *the good* of the person, the image of God, by protecting his *goods*." In this perspective, as John Paul II observes, the negative precepts of the Decalogue – which themselves are rooted in the commandment to love our neighbor as ourselves – "express with particular force the ever urgent need to protect human life, the communion of marriage," and other fundamental goods (no. 13).[29]

Without pursuing this point any further, there is one other problem I would like to note with regard to O'Rourke's analysis and that is his use of the equivocal term "subordinate" (sometimes he and Ashley speak of "sacrificing" a good). Even according to their own ethical system (i.e., "prudential personalism") that he and Ashley have worked out, one is never morally permitted to directly take the life of an innocent human person (and recall that both men grant that the PVS patient is a human person) or to "directly contradict" any basic human good. Human acts that do so are "intrinsically evil" (See Ashley, 1994, p. 92).[30] "Subordinate," then, in this context, can never be taken to mean, as morally licit, to "directly kill the innocent."

By speaking in terms of "subordinating" goods, however, O'Rourke begs the question of *when* it is morally allowable to withhold or withdraw tubal feeding from a PVS patient. That is, to say that one good, of a so-called lower order (bodily life, physiological functioning, physical existence) is subordinate to another good, of a so-called higher order (truth, contemplation, the ultimate end which is the "spiritual end of human life") says nothing about the *criteria* by which one can judge whether one's subordination or sacrifice of a good is morally upright (i.e., as in letting the patient die a natural death) or morally wrong (i.e., as in intentionally killing the patient) in a particular situation.[31] It is enough to say too, that John Paul II, Grisez, May, myself, and others who defend feeding PVS patients do not "absolutize" the good of *bodily life* as O'Rourke, Charles Curran, Thomas Shannon, and others claim, but rather we absolutize the *moral norm* that states human persons are never to directly kill other innocent human persons even if for a good end.[32]

## 12.2.2 O'Rourke and the March 2004 Papal Address

Interestingly enough, in his April 2005 Detroit debate with May, contrary to his previous argument that such "medical therapy" as tube feeding "does not offer 'hope of benefit' to the [PVS] patient" (O'Rourke, 1988, part 1, p. 4.),[33] O'Rourke seemed willing to concede the point, in light of John Paul II's address, that there *is* "hope of benefit."[34] He then went on to emphasize however, that the "heavy human, psychological and financial burden" associated with caring for patients in a PVS leads to the conclusion that one need not feed them.[35] In religious and secular news reports on the Schiavo case, O'Rourke was also quoted as saying that one could specify in an advance medical directive that one would not want a feeding tube inserted should one fall into a PVS in order to spare the family the burden of caring for the patient.[36]

William E. May has responded well to the "excessive burden" argument by noting that the total cost of caring for PVS patients could be quite high, but that "in our affluent society, which provides such care for other persons in severely debilitated conditions [e.g., quadriplegics, multiple amputees, and the severely mentally and physically handicapped], it would be unfair to deprive the permanently unconscious of their fair share of such care." "[I]f one were to seek to avoid the expense involved in caring for such persons by denying them food," May continues, "the means chosen to avoid this expense would not be the withholding of food as such but rather the subsequent death of these persons" (May, 1998/1999, part 1, p. 4).[37] Because tubal feeding and hydrating of PVS patients is generally neither useless nor overly burdensome, it ought to be given unless, for example, the health care resources of one's family and/or society were truly impoverished and needed to be expended on other sick patients, the patient's body were unable to assimilate the food, he or she were immanently dying, or the tube was causing a serious infection (in the latter case, I take it, one would still try to hand feed the patient).[38]

John Finnis also treats the question of what constitutes "excessive expenditure," showing how families do not have to bankrupt themselves in caring for family members in a PVS. Finnis recognizes that although human life "even in irreversible unconsciousness is of intrinsic value, and may not be intentionally destroyed by act or omission, that value wholly lacks the further goods which normally accompany it (knowledge, friendship, play and skill, communication in prayer, etc)." Thus, he asks, might not someone contemplating being in so radically deprived a state reasonably decide that any use of hospital and specifically medical resources would be excessive? "If so," Finnis argues, the patient "could judge that the duty to give and accept ordinary care requires no more than this: the giving of such food, water and nursing care as can be provided from the resources available in one's home" (Finnis, 1994, p. 176).

Against the position of those theologians such as O'Rourke, who would deny PVS patients food and water,[39] John Paul II teaches in the relevant moral part of his March 2004 address on the vegetative state:

> The sick person in a vegetative state, awaiting recovery or a natural end, still has the right to basic health care (nutrition, hydration, cleanliness, warmth, etc.), and to the prevention of complications related to his confinement to bed. He also has the right to appropriate rehabilitative care and to be monitored for clinical signs of eventual recovery (2004, no. 4).

The Pope continues:

> I should like particularly to underline how the administration of water and food, even when provided by artificial means, always represents a *natural means* of preserving life, not a *medical act*. Its use, furthermore, should be considered, in principle, *ordinary* and *proportionate*, and as such morally obligatory, insofar as and until it is seen to have attained its proper finality, which in the present case consists in providing nourishment to the patient and alleviation of his suffering (Ibid.).

Thus, John Paul concludes, even if hope for recovery fades when the vegetative state is prolonged beyond a year, this:

> cannot ethically justify the cessation or interruption of *minimal care* for the patient, including nutrition and hydration. Death by starvation or dehydration is, in fact, the only possible outcome as a result of their withdrawal. In this sense it ends up becoming, if done knowingly and willingly, true and proper euthanasia by omission (Ibid.).

Everything that I have argued in this article I believe to be compatible with the March 2004 address of the late Holy Father. Pope John Paul II has now spoken authoritatively on an issue that had been vigorously debated for over twenty years in the Catholic Church, and in doing so has given Catholics (and others) invaluable moral guidance for properly forming their consciences on such a difficult but vital end-of-life matter. He is not teaching (as some do in fact teach), let it be noted, that artificially assisted nutrition and hydration must be provided in *every* case (note the Pope's use of the words "in principle"), nor is he insensitive to the difficulties faced by families who care for PVS patients (note how he speaks of "support" for families and his suggestions for how to do this in no. 6). John Paul II is concerned, however, that PVS patients always be treated with respect, in keeping with the fact that they remain human persons despite their debilitated condition.

John Paul II has provided the faithful with a magisterial document that is not only consistent with the tradition, but one that is both fully informed by the current medical facts and sure to be pastorally helpful in real life health care situations.[40] May Fr. O'Rourke, who has done so much good for the Catholic Church in bioethics, reread this papal document and then seriously reconsider his own view. Doing so, I believe, would be of great benefit to future patients, such as Terri Schiavo, who might find themselves severely brain damaged or in a PVS.[41]

# Notes

[1] The literature on the medical and moral aspects of the PVS is enormous and highly controversial. Ashley and O'Rourke accurately observe that patients in a PVS "show signs of arousal and may grimace or make other reflex movements, but show no signs of awareness or response to communication." This condition is distinguished from those such as the "locked-in state" and "coma" (See Ashley and O'Rourke, 1997, p. 423). For contrasting medical views on the PVS, see e.g., Jenett (1997, pp. 169–188), see especially pp. 170–174 and Shewmon (1997, 30–96), see especially 58–60. Jennett argues that medically assisted food and water can be withdrawn from PVS patients. Shewmon powerfully challenges many of the widely held assumptions about the PVS on medical and metaphysical levels.

[2] This two and a half-hour debate took place for the priests of the Archdiocese of Detroit on April 26, 2005. O'Rourke's presentation consisted of Power Point print out slides that he handed out. May's presentation was a paper titled, "Caring for Persons in the 'Persistent Vegetative State' and Pope John Paul II's March 2004 Address 'On Life-Sustaining Treatments and the Vegetative State,' " *Medicina e Morale*, 2005. May and O'Rourke have debated before in published articles, e.g., see May (1998/1999/) and O'Rourke [1988].

[3] John Paul II's address to the participants at a congress on the "vegetative state" is titled, "Life-Sustaining Treatments and the Vegetative State," 2004 (see references). For O'Rourke's critical

reaction, see, e.g., Graham [2005]. For a mere sampling of the other reactions to this address, see Hamel and Panicola (2004, pp. 6–13); Cahill (2005, pp. 14–17) (unfavorable); Doerflinger (2004, pp. 2–4); Cole (2004, pp. 62–66) and Ford, 2005, pp. 3–4 (favorable). See further Kopaczynski (2004, pp. 473–482) and the lengthy "Colloquy Section" of *The National Catholic Bioethics Quarterly* 4.3 (Autumn 2004): 447–459 with various authors commenting on the papal address.

[4] For a recent critique of Ashley and O'Rourke on the PVS published just before the papal address, see Torchia (2003, 719–730) and the exchange between those authors and Torchia in *The National Catholic Bioethics Quarterly* 4.2 (Summer 2004): 1–2.

[5] Pope Pius XII, "Prolongation of Life," as quoted in Ashley and O'Rourke (1997, p. 425), their emphasis. An excerpt of this address can be found in O'Rourke and Boyle (Eds.), *Medical Ethics: Sources of Catholic Teachings*, Third Edition, pp. 280–281.

[6] Here they cite O'Rourke and DeBlois, (1992, p. 20–28). In Ashley and O'Rourke (2002), p. 195, explicit appeal is made to St. Thomas Aquinas, *Summa theologiae*, I–II, q. 1, a. 1 to support the idea that it is by means of human acts that human persons are able to pursue the spiritual purpose of life.

[7] See also, Ashley and O'Rourke (2003), p. 193. Cf. Matthew 6:33.

[8] Emphasis added. Note the reference here to, among other authors, the often cited and influential article by O'Rourke (1986, pp. 321–323, 331). I want to note here as others have (e.g., Gilbert Meilander) that it is not the *food* that is artificial but the *means* of providing it to the patient. Although admittedly the food is not "pizza in a tube," to speak of it being "artificial" runs the danger of biasing the discussion from the beginning in favor of withholding or withdrawing tube feeding.

[9] Ashley and O'Rourke (2002), maintain this view in their most recent joint writing on the issue (pp. 189–194, 195–196). See also O'Rourke and Norris (2001, pp. 201–217). In O'Rourke and DeBlois (1995, p. 24–27), the authors claim: "Prolonging the life of persons in PVS does not seem to enhance their ability to strive for the purpose and goods of life" (p. 27).

[10] Ashley and O'Rourke prefer the older terms *ordinary* and *extraordinary* rather than the newer terms *proportionate* and *disproportionate*, but "define them as does the *Declaration on Euthanasia, 1980*, by the proportion of benefit to burden, while at the same time maintaining, contrary to proportionalism, the principle that *direct* killing is intrinsically and always unethical, although 'letting die' when the therapy will not benefit the patient (indirect killing in accord with the principle of double effect) is ethically justifiable" (Ashley and O'Rourke, 1997, pp. 420–421).

[11] "[C]lassical moral theology," they add, "held that taking food and drink even in a normal manner ceases to be obligatory for a patient if its benefit no longer exceeds its burden" (Ashley and O'Rourke, 1997, p. 421). Thus, Mackler (*Introduction to Jewish and Catholic Bioethics*) has perceptively noted that Ashley and O'Rourke are among those theologians who work with a more expansive notion of burden and benefit than many others would be willing to accept, or if they do accept it, would not extend it to the provision of artificially given food and drink. He summarizes their view as follows: Ashley and O'Rourke "understand the key issue in treatment decisions to be the 'proportion of benefits to burdens'; any form of care – not only medical treatment – may be foregone if it is disproportionately burdensome. Treatment may be foregone whenever it imposes a 'grave burden' or fails to be 'truly beneficial,' even for patients who are not imminently dying. Burdens to be considered include not only those directly entailed by treatment but also 'the burdens of total care.' For permanently unconscious patients, burdens include 'the indignity of existing in a state of persistent cognitive-affective deprivation [which] can be counted as a serious burden to the patient, even if the patient is unconscious,' as well as burdens experienced by the family and caregivers" (p. 102). Here Mackler cites Ashley and O'Rourke (1997, pp. 421–423) (but it also includes p. 424).

[12] On p. 195 of *Ethics of Health Care* Ashley and O'Rourke speak ambiguously about PVS and a fatal pathology. On the one hand, they seem to say that PVS patients presently suffer from a fatal pathology; on the other hand, they then say that patients in this condition should not be killed, "but *when they contract a fatal pathology* [do they not already have one according to the authors?],

overcoming or resisting that pathology would usually involve extraordinary or disproportionate means. Thus, if a person with PVS or advanced Alzheimer's disease is unable to eat without artificial hydration and nutrition, it is reasonable to maintain that such therapy need not be utilized" (emphasis added).

[13] This argument is shared by the Texas Conference of Catholic Bishops in their statement, "On Withholding Artificial Nutrition and Hydration," in *Origins* 20 (1990): 53–55.

[14] In addition to Berkman's examination of the central arguments in the debate (see 89–95), his article has a fascinating overview of the medical literature past and present on tube feeding (see 73–79, 96–99). Over the last twenty years U.S. courts and medical societies have held, unfortunately, that artificially assisted food and water are medical treatment that can be removed from PVS patients. Although recent studies have given mixed reviews on the therapeutic value of tube feeding, this fact would not lead to the conclusion that PVS patients should not be fed. An admittedly time-consuming alternative could be spoon feeding. See Bro. Perkins (2004, pp. 1–4).

[15] See e.g., May et al., 1987, pp. 203–217. See also (Diamond, 1987, pp. 73–76).

[16] See part two of O'Rourke's 1988 response to May, where O'Rourke claims it is "irrelevant to state that PVS is not a fatal pathology *after* life support has been utilized" (p. 4).

[17] I believe that the study Diamond is referring to is Andrews, 1996, pp. 13–16. This study calls into question Ashley and O'Rourke's confidence in the medical profession's ability to accurately diagnose the PVS condition.

[18] See also pp. 232–235, 239–259. See further Gormally (1994, pp. 141–143).

[19] See Ashley and O'Rourke (1997, p. 422).

[20] See also Gormally (Ed.), 1994, pp. 142–143, footnote 11.

[21] See Keown, 2002, p. 220. Keown raises the question in the context of the famous Bland case.

[22] For a cogent argument against dualism, see e.g., Grisez (1977, pp. 323–330 and 377–379) and Grisez and Boyle, 1979, pp. 377–379.

[23] In fact, O'Rourke grants that PVS patients are human beings (see O'Rourke, 1998, part 1, p. 3). Without committing the error of treating a person's life instrumentally, i.e., as having value only insofar as it is useful in achieving further conscious purposes, one can admit that PVS patients are no longer experiencing or participating in the basic good of human life to its fullest. But, it must be said, a "person's life cannot simply be instrumental to other goods of that person. For a person's life is not something other than his or her very self, and so the living organism, however deprived, cannot be separated from the person in the way an instrument can be separated from the purposes for which it is used" (Boyle, 1997, p. 192). Life has an intrinsic value that transcends its being a condition for the realization of other goods the person wants to achieve. Moreover, no bad condition can lessen a person's human dignity. Boyle also shows how PVS patients can be harmed or benefited even when they do not actually experience the harm or benefit (see Ibid. pp. 192–194).

[24] See e.g. Ashley and O'Rourke (1997, pp. 31–34), where they express their long-standing disagreement with dualistic anthropologies. They hold that their anthropology is non-dualistic – emphasizing the unity of the person, body and soul – and inspired by the Aristotelian and Thomistic tradition. See also pp. 3–21.

[25] During the Detroit debate, O'Rourke said that he likes to speak of "quality of function," rather than "quality of life" because of the ambiguity of the latter term. Mackler says that in contrast to Richard McCormick, S.J., "[t]he extent to which Ashley and O'Rourke's avoidance of the term ['quality of life'] reflects their concerns about possible abuse of the term and the extent to which it simply reflects deference to magisterial sensitivities are unclear" (Mackler, 2003, p. 119, footnote 73).

[26] O'Rourke quotes Ashley: "[t]he human body is human precisely because it is a body made for and used by intelligence. Why should it be dualism to unify the human body by subordinating the goods of the body to the good of the immaterial free and contemplative intelligence?" (O'Rourke, 1988, part 2, p. 3) But even if, for the sake of the argument, one grants this point, it does not, as we will see, determine whether the "subordinating" is morally good or morally evil. That still needs to be rationally determined.

[27] See e.g., Finnis and Grisez, 1981, p. 21–31, especially 28–31. Moreover, as Finnis has argued, a "first-order good" (e.g., life) may not, according to St. Thomas, be intentionally harmed in order to save a "third-order good" (e.g., friendship with God). See John Finnis, 1980, pp. 94–95 (Finnis cites *Summa theologiae*, I–II, q. 64, a. 5 and 6).

[28] On this, see May (2004, pp. 128–129).

[29] The Pope notes that the command that we are to love our neighbor as ourselves expresses "the singular dignity of the human person, the 'only creature that God has wanted for its own sake' " (*Veritatis splendor*, no. 13; the internal citation is from *Gaudium et spes*, no. 22).

[30] See Ashley, 1994, p. 92. This is why Ashley and O'Rourke's teleological moral system is not, as we have seen, of the proportionalist kind. It does commendably uphold and defend moral absolutes.

[31] Ashley argues that bodily life, as "the least in the hierarchy of basic goods . . . can be sacrificed to the higher goods, but only on condition that the sacrificial act (1) does not involve an injustice; (2) is not intrinsically evil" (Ashley, 1994, p. 92). But one needs some objective moral standard in order to show precisely *why* the sacrifice is not unjust or intrinsically evil. What Ashley says simply begs the question. I argue that denying food and water to PVS patients *is* unjust.

[32] Charles E. Curran has accused John Paul II of making human bodily life an "absolute" good, and he appeals to the *same* text of Pius XII that Ashley and O'Rourke do (see Curran, 2005, pp. 114–115; 124, footnote 19. Curran mistakenly says the date for the Pope's address is November, 24 1951, when in actual fact the year is 1957). For Shannon's view, see Shannon, 2005, p. 19. We obviously would recognize that there are many situations where one is under no moral obligation to preserve human life – one's own or others – at all costs. Contrary to O'Rourke's assertion about absolutizing bodily life, then, life need not be prolonged indefinitely when, e.g., to do so would involve extraordinary means. Thus, I would grant that even the factor of *cost* in relation to what care givers are actually capable of doing and affording, could justify withdrawing or withholding a feeding tube from someone in a PVS without the intention to kill. Also, there is a legitimate sense in which Christians can speak of "sacrificing" certain earthly goods (e.g., one's life) to "higher goods" (e.g., Christian faith) as in martyrdom. But the martyr does not intend his or her death, but freely accepts its loss as a side effect (*praeter intentionem*) of being faithful to God.
See also the important article by Cataldo (2004, pp. 513–536) on "[h]ow subordination to the last end [of man] does not eliminate the subordinate duty to preserve life for Aquinas . . ." (p. 524, footnote 37); see pp. 530–536. Cataldo insightfully shows, against interpretations such as O'Rourke's, how Pius XII's address is directed to the patient "who is conscious and functionally able to pursue the spiritual ends of life . . . If the type of patient to whom Pius XII refers was a permanent-coma patient, then the problem he discusses of attaining this higher good would not be relevant. It is precisely because the patient in question is actually able to attain the spiritual good of life that Pius XII's discussion of burdensome means makes sense. The statement is a narrow delimitation of the moral obligations of a person who has the awareness necessary to engage the spiritual goods of life; it is not about a broad standard for the meaning of benefit or beneficial means for both conscious and unconscious patients alike" (p. 535).

[33] Although I disagree with this position (attractive as it might be) I do not have the space here to respond to it. Germain Grisez and Joseph Boyle, however, accept this position that was first articulated by Grisez in the mid 1980's, as long as the intention is not to commit suicide or help one's loved ones to do so, but rather to spare the caregivers the extreme burdens of care.

[34] This concession seemed strained in the sense that O'Rourke was also highly critical of the address in his debate with May. Yet he said that he accepted it as non-definitive "Church teaching" (see also his comments in Graham, 2005). It was not clear from O'Rourke's Power Point print-out and oral remarks, however, that he had, in fact, abandoned his "spiritual end of life" argument (e.g., on his print-out he gave the same interpretation of Pius XII's 1957 address he has given over the years and he quoted the same passage from Ashley that appears in footnote 26 above), despite his apparent partial agreement with the Pope, who he said, had accepted the arguments of the Grisez and May position that tube feeding is comfort care and that it is a benefit to keep PVS patients alive. Thus, I seriously doubt whether O'Rourke would counsel someone faced with the question of tube feeding a family member who is in a PVS to follow the Pope's moral teaching in the address.

[35] May and O'Rourke debate in Detroit, April 26, 2005.

[36] See Bono [2005].

[37] May is summarizing the conclusions of May, et al., 1987. The authors of that essay had acknowledged that "*if it is really useless or excessively burdensome* to provide someone with nutrition and hydration, then these means may rightly be withheld or withdrawn, *provided* that this omission does not carry out a proposal to end the person's life but rather is chosen to avoid the useless effort or the excessive burden of continuing to provide the food and fluids" (May, et al., 1987, p. 209). But because they are *generally* a useful means in preserving the great good of human bodily life and can be provided without excessive burden, "both morality and law should recognize a strong presumption in favor of their use" (Ibid. p. 211). See also the Pennsylvania Conference of Catholic Bishops, "Nutrition and Hydration: Moral Considerations," in *Origins* 21 (1992): 542–553 and the U.S. Bishops' Pro-Life Committee, "Nutrition and Hydration: Moral and Pastoral Reflections," in *Origins* 21 (1992): 705–711 who offer similar perspectives.

[38] For an excellent defense of the position that PVS patients should be fed and hydrated in most cases, unless it is contraindicated as truly futile or excessively burdensome, see Ryan [2000]. See also Boyle (1997, pp. 189–199). Boyle argues that, "the relevant moral rule is that we must do what we can to care for patients in PVS, and to maintain ties with them which show our respect for their human dignity [Boyle, Grisez and others speak of maintaining "solidarity"]. But the application of this rule to the situation of different people with different resources and varying forms of social support must inevitably lead to different concrete moral judgments as to what it is right for them to do" (p. 199). If Boyle were writing this essay today with an eye to the Pope's March 2004 address, I wonder if some of his arguments and conclusions would be revised. For Grisez's most recent position, see Grisez (1997a; 1997b), Question 47: "May a husband consent to stopping feeding his permanently unconscious wife?" pp. 218–225. Many of the footnotes of this question respond to O'Rourke.

[39] Among other reasons for not feeding PVS patients, Ashley and O'Rourke fear that doing so will appear to the public as "vitalism" and thus create a climate favorable to the legalization of euthanasia (see Ashley and O'Rourke, 1997, p. 428). I have found that most of their writing on this issue expresses this concern. However, we already have a climate that is in many ways favorable to euthanasia, as John Paul II has noted in *Evangelium vitae,* 1995, where he describes affluent Western societies as having a "culture of death." One could argue that the unwillingness to feed and hydrate an allegedly PVS patient such as Terri Schiavo is a *symptom* of this anti-life attitude which has been growing for many years now. By not feeding PVS patients, we only contribute to the false notion that they have lives that are not worth living. During the Detroit debate, O'Rourke also made much of the fact that people have often told him that they would not want to be tube fed if they were in a PVS because they would not want to live in such a debilitated condition.

[40] See on this matter the insightful article by Garcia [2006].

[41] See also on the Schiavo case, Stith [2005] and Furton [2005]. For a defense of O'Rourke's position, see Eberl [2005]. As this paper was going to press in late 2006, there were a number of articles reporting on patients who had seemed to either regain consciousness or partial consciousness after a diagnosis of PVS. There were also reports from England of tests that were done on a patient said to be in a PVS, using functional magnetic resonance imaging, which indicated that her brain activity was identical to that of healthy brains that had not suffered trauma. See Balch [2006].

# References

Andrews, K. et al. (1996). Misdiagnosis of the vegetative state: retrospective study in a rehabilitation unit. *British Medical Journal, 313,* 13–16.

Ashley, B. M. (1994). What is the end of the human person?: The vision of God and integral human fulfillment. In L. Gormally (Ed.), *Moral truth and moral tradition: Essays in honor Of Peter Geach and Elizabeth Anscombe.* Dublin: Four Courts Press.

Ashley, B. M., & O'Rourke, K. D. (1997). *Health care ethics: A theological analysis* (4th ed.). Washington, DC: Georgetown University Press.

Ashley, B. M. & O'Rourke, K. D. (2002). *Ethics of health care: An introductory textbook*, (3rd ed.). Washington, DC: Georgetown University Press.

Balch, B. (2006, October). Breakthrough research shows consciousness in PVS patient. *National Right to Life News, 33*, 20.

Berkman, J. (2004). Medically assisted nutrition and hydration in medicine and moral theology: A contextualization of its past and a direction for its future. *The Thomist, 68*, 69–104.

Bono, A. (2005). *Schiavo Case Raises Issue of Advance Medical Directives*. [On-line] Available at: http://www.the-tidings.com/2005/0401/schiavo.htm.

Boyle, J. (1997). A case for sometimes tube-feeding patients in persistent vegetative state. In J. Keown (Ed.), *Euthanasia examined: ethical, clinical and legal prespectives*, (pp. 189–199). Cambridge: Cambridge University Press.

Boyle, J., & Germain, G. (1979). *Life and death with liberty and justice: A contribution to the euthanasia debate*. Notre Dame, IN: University of Notre Dame Press.

Cahill, L. S. (2005, April 25). Catholicism, death and modern medicine. *America, 195*, 14–17.

Cataldo, P. (2004). Pope John Paul II on nutrition and hydration: A change of Catholic teaching? *National Catholic Bioethics Quarterly, 4*(3), 513–536.

Cole, B. (2004, December). Why so long to make a decision? *Homiletic & Pastoral Review, 105*, 62–66.

Colloquy. (2004). Colloquy section. *The National Catholic Bioethics Quarterly, 4*(3), 447–459.

Congregation for the Doctrine of the Faith. (1980). *Declaration on Euthanasia* [On-line] Available at: http://www.vatican.va/roman_curia/congregations/cfaith/documents/rc_con_cfaith_doc_19800505_euthanasia_en.html

Curran, C. E. (2005). *The moral theology of Pope John Paul II*. Washington, DC: Georgetown University Press.

Diamond, E. F. (1987). The AMA statement on tube feeding. *The Linacre Quarterly, 54*(2), 73–76.

Diamond, E. F. (1999). Medical issues when discontinuing ANH: Further debate on feeding the 'vegetative'. *Ethics & Medics, 24*(9), 1–2.

Doerflinger, R. (2004). John Paul II and the 'vegetative' state. *Ethics & Medics, 29*(6), 2–4.

Eberl, J. (2005). Extraordinary care and the spiritual goal of life. *The National Catholic Bioethics Quarterly, 5*(3), 491–501.

Finnis, J. (1980). *Natural law and natural rights*. Oxford: Oxford University Press.

Finnis, J. (1994). 'Living will' legislation. In L. Gormally (Ed.), *Euthanasia, clinical practice and the law* (pp. 167–176). London: Linacre Centre for Health Care Ethics.

Finnis, J. & Grisez, G.(1981). The Basic principles of natural law: A reply to Ralph McInerny. *American Journal of Jurisprudence, 25*, 21–31.

Ford, N. M. (2005). Thoughts on the papal address & MANH. *Ethics & Medics, 30*(2), 3–4.

Furton, E. (2005). On the death of Terri Schiavo. *Ethics & Medics, 30*, 3–4.

Garcia, J. L. A. (2006). A Catholic perspective on the ethics of artificially providing food and water. *Linacre Quarterly, 73*(2), 132–152.

Gormally, L. (1994). *Euthanasia, clinical practice and the law*. London: The Linacre Centre.

Graham, J. (2005) *Schiavo case put priest on hot seat* [On-line] Available at: http://www.chicagotribune.com/news/local/chi-0504240028apr24,1,3428290.story?coll=chi-news-hed&ctrack=2&cset=true.

Grisez, G. (1977). Dualism and the New morality. In *Atti del congresso internazzionale Tomasso d'Aquino nel suo settimo centenarrio, L'agire morale*, (Vol. 5, pp. 323–330). Naples: Edizioni Domenicane Italiane.

Grisez, G. (1997). *The way of the lord Jesus, Vol. 3: Difficult moral questions*. Quincy, IL: Franciscan Press.

Hamel, R. & Panicola, M. (2004, April 19). Must we preserve life? *America*, 6–13.

Jenett, B. (1997). Letting vegetative patients die. In J. Keown (Ed.), *Ethical, clinical and legal perspectives* (pp. 169–188). Cambridge: Cambridge University Press.

John Paul II. (1993). *Veritatis Splendor.* Available at: http://www.vatican.va/holy_father/john_paul_ ii/encyclicals/documents/hf_jp-ii_enc_06081993_veritatis-splendor_en.html.

John Paul II. (1995). *Evangelium Vitae* Available at: http://www.vatican.va/roman_curia/ congregations/cfaith/documents/rc_con_cfaith_doc_19800505_euthanasia_en.html.

John Paul II. (2004). *Life-sustaining treatments and the vegetative state.* Available at: http://www.vatican.va/holy_father/john_paul_ii/speeches/2004/march/documents/hf_jp-ii_ spe_20040320_congress-fiamc_en.html.

Keown, J. (2002). *Euthanasia and public policy: An argument against legalization.* Cambridge: Cambridge University Press.

Kopaczynski, G. (2004). Initial reactions to the Pope's March 20, 2004, allocution. *The National Catholic Bioethics Quarterly, 4*(3), 473–482.

Mackler, A. L. (2003). *Introduction to Jewish and Catholic bioethics.* Washington, DC: Georgetown University Press.

May, W. E. (1998/1999). Tube feeding and the 'vegetative' state: The case for artificial feeding. *Ethics & Medics, 23*(12)/24(1), 1–2/1–2.

May, W. E. (2000). *Catholic bioethics and the gift of human life.* Huntington, Indiana: Our Sunday Visitor Press.

May, W. E. (2004). Contemporary perspectives on Thomistic natural law. In J. Goyotte, M. Latkovic, & R. Myers (Eds.), *St. Thomas Aquinas and the natural law tradition: contemporary perspectives.* Washington, DC: The Catholic University of America Press.

May, W. E. (2005). Caring for persons in the 'persistent vegetative state' and Pope John Paul II's March 2004 address 'on life-sustaining treatments and the vegetative state. *Medicina e Morale, 55* (Maggio/Giungno), 535–555.

May, W. E. et al. (1987). Feeding and hydrating the permanently unconscious and other vulnerable persons. *Issues in Law in Medicine, 3,* 203–217.

O'Rourke, K. D. (1986, November 22). The A.M.A. statement on tube feeding: An ethical analysis. *America, 155,* 321–323, 331.

O'Rourke, K. D. (1988). Evolution of church teaching on prolonging life. *Health Progress, 69*(1), 28–35.

O'Rourke, K. D. (1998). On the care of 'vegetative' patients: A response to William E. May. *Ethics & Medics, 24*(4), 3–4.

O'Rourke, K. D., & DeBlois, J. (1992). Removing life support: motivations, obligations. *Health Progress, 73*(4), 20–28.

O'Rourke, K. D. & DeBlois, J. (1995). Issues at the end of life. *Health Progress, 76*(6), 24–27.

O'Rourke, K. D. & Norris, P. (2001). Care of PVS Patients: catholic opinion in the United States. *Linacre Quarterly, 68*(3), 201–217.

Pennsylvania Conference of Catholic Bishops. (1992). Nutrition and Hydration: Moral Considerations. *Origins, 21,* 542–553.

Perkins, I. (2004). Feed the sick, nourish the person: On using hand feeding instead of tube feeding. *Ethics & Medics, 29*(9), 1–4.

Pius XII. (1957). Prolongation of life. See Ashley,*Health Care Ethics,* p. 425; or in (1999) O'Rourke & Boyle (Eds.), *Medical Ethics: Sources of Catholic Teachings,* (3rd ed. pp. 280–281). Washington, DC: Georgetown University Press.

Ryan, P. (2000).*The value of life and its bearing on three issues of medical ethics.* Available at: http://www.wf-f.org/Sum2K-Ryan.html.

Shannon (2005, June 6). The legacy of the Schiavo case. *America, 195,* 17–19.

Shewmon, D. A. (1997). Recovery from 'brian death': A neurologist's apologia. *TheLinacre Quarterly, 64*(1), 30–96.

Stith, R. (2005). Death by hunger and thirst. *Ethics & Medics, 30,* 1–2.

Texas Conference of Catholic Bishops. (1990). On withholding artificial nutrition and hydration. *Origins, 20,* 53–55.

Torchia, J. (2003). Artificial hydration and nutrition for the PVS Patient: Ordinary care or extraordinary intervention. *The National Catholic Bioethics Quarterly, 3*(4), 719–730.

Torchia, A. and O'Rourke, K. in the Colloquy section (2004). *The National Catholic Bioethics Quarterly, 4*(2), 1–2.

United States Conference of Catholic Bishops. (1992). Nutrition and hydration: Moral and pastoral reflections. *Origins, 21,* 705–711.

United States Conference of Catholic Bishops. (1994). Ethical and religious directives for catholic health services. *Origins, 24,* 449–464.

# Part IV
# Concluding Reflections

# Chapter 13
# Ten Errors Regarding End of Life Issues, and Especially Artificial Nutrition and Hydration

Christopher Tollefsen

Recent events, including the conflict over Terri Schiavo, and the death of Pope John Paul II, have made necessary renewed attention to the Church's teaching regarding end of life care and treatment, especially the artificial provision of nutrition and hydration to those unable to feed themselves.

According to the Church's general approach to end of life matters, it is not morally incumbent upon a patient to accept all forms of care or therapy that may be offered. Pope Pius XII, in his 1958 "Address to the First International Congress of Anesthesiologists," articulated the following principles. First, one is obliged to use only "ordinary" means, or forms of treatment; "extraordinary" forms of treatment may be refused. Further, the primary right to refuse treatment rests in the patient herself; it is not for doctors or family members to decide for an otherwise competent patient, what forms of treatment should be accepted. However, in some cases, decisions regarding patients must be made by proxy, preferably by a family member with the patient's best interests and will at heart. Yet, finally, it is also permissible for the family to remove treatment for an incompetent patient if the burdens, such as the cost, of the treatment grow too great for the family (Pope Pius XII, 1958).

Pope John Paul II directed attention to this teaching in the context of his remarks on the artificial provision of nutrition and hydration to those in permanent (or persistent) vegetative states. The Pope declared that the provision of food and water should "in principle" be considered care, not treatment, and ordinary care at that. Nor should patients in PVS be considered truly to be "vegetables" or to be living some kind of sub-human, vegetative, existence (John Paul II, 2004). The Pope's insistence that such patients continue to be provided with food and water took many by surprise, in part because some hospitals seem to have developed the regular practice of withdrawing ANH from such patients, and in part because the Pope seemed to some to be making the very strong claim that ANH could *never* be withdrawn from any patient. In consequence, much commentary has been generated over the Pope's remarks, especially in context of discussion of the Terri Schiavo case.

Christopher Tollefsen
University of South Carolina
Email: Tollefsen@sc.ed

C. Tollefsen (ed.), *Artificial Nutrition and Hydration:*
*The New Catholic Debate*, 213–226. © Springer 2008

These are difficult cases, to be sure. Moreover, discussion of care and treatment, of ordinary and extraordinary, of proportionate and disproportionate, quickly gets us into discussion of intention, side effect, double effect, and so on. These are murky waters, to which even highly trained philosophers and theologians have not always brought clarity. In this paper, my aim is to work through some of the issues associated with these cases by bringing to light what I would consider to be several more or less serious errors in their treatment. Some of these errors are, I think, quite common; others might merely be hypothetical. Some are merely errors in the articulation of otherwise sound positions. Some are conceptual confusions, arising from failure to understand entirely certain elements of action theory. And some are mistakes of moral reasoning, of the application of moral norms to the cases at hand. I do not, for the most part, identify sources for these errors, when there are any; the exercise in this paper is intended more as a matter of moral reasoning and conceptual clarification than of diagnosis of error or failure.

## 13.1 It is Permissible to Act with the Intention of Letting Someone Die

The first error on my list makes a simple, but common error about the distinction between killing, and allowing to die.

According to constant Church teaching, it is, at least, impermissible ever for a private citizen to act with the intention of killing another person. The qualification "private citizen" is necessary, for the Church has, in the past, held that public officials are permitted intentionally to kill malefactors. They are not allowed to intentionally kill the innocent; however, as private citizens are not allowed intentionally to kill anyone, it is not necessary to include the qualification "innocent" (see Grisez, 1970; Boyle, 1989; Brugger, 2003).

Private citizens may, however, defend themselves against wrongful attack, even if, in the course of their self-defense, they must inflict lethal injury. But, following Thomas Aquinas' analysis, the Church has held that in such cases, agents do not, if acting permissibly, act with the intention of killing, but rather with the intention of using force to repel an attack. What damage, including death, is done to the attacker is a side effect of the agent's action, not intended, but only accepted (Aquinas, 1996, I–II, 64, a. 7).

It is this analysis of the difference between what is intended, as a means or as an end, and what is accepted as a side effect, that makes possible the Church teaching that agents are not under an obligation to accept all forms of life saving or preserving treatment. For if the distinction were effaced, such that there was no difference between accepting a side effect and intending, then any refusal that had as its consequence the death of the patient would amount to self-killing (and any similar decision by a proxy would amount to murder). But patients may refuse treatment for many reasons – because of the expense, say, or because of the side effects of the treatment itself. In rightly rejecting treatment, however, agents must be acting (refusing) so as to obtain some benefit (limited expense) such that death is not intended either as a means or an end. It is, rather, accepted as a side effect.

Similarly, proxies may decide to refuse treatment on behalf of an incompetent agent for the sake of some legitimate end, accepting death as a side effect, but they may not refuse treatment in order to bring about death, nor may they in any way intend the death of the patient. But these distinctions are muddled by saying that an agent, whether the patient or a proxy, can act *with the intention* of *allowing the death* of the patient. For the whole purpose of the intend-foresee distinction was to make the point that the permitted effects were not part of the agent's intention at all. Joseph Boyle has suggested the most plausible reason for this focus on intention: in a moral system in which certain kinds of willed actions are to be absolutely and without exception prohibited because of the damage that they do to human goods, the prohibition cannot extend to any relationship whatsoever between the will and the damage, because in virtually everything an agent does, some good is damaged, at least in virtue of its not being pursued. If, then, there are to be moral absolutes of the sort articulated constantly throughout the Catholic tradition, they must govern only what is intended, for it is always in an agent's power not to intentionally harm basic goods (Boyle, 1989).

The proper way to express what is permitted, then, is to say that agents may refuse treatment and accept death as a side effect (similarly, they may accept painkillers, and accept death as a side effect). But they may not do anything, whether refusing treatment, or utilizing painkillers, with the intention of accepting death.

## 13.2 The Relevant Distinction is Between Acting and Omitting

Focusing on cases in which patients or their proxies refuse some offered form of treatment, one might be tempted to conclude that what is permissible is omitting something, where the omitting leads to death, whereas what is impermissible is *acting* in some way that leads to death. One might then be puzzled at the Pope's insistence that patients in PVS be fed; surely, if omission is permissible as such, then omitting to feed is permissible.

The distinction between intending and accepting as a side effect is, however, manifestly not the same as the distinction between acting and omitting. Furthermore, I would argue, it is only the former distinction that is of widespread importance in ethics. There are two parts to my defense of this claim.

The first part is simply to point out that much omitting is intentional, and that wrongful acts may intentionally be done precisely by omitting. If I desire the inheritance, I might omit to give my uncle his medicine, precisely in order to bring about his death; and then I surely include his death as part of my intention. Of course, not all omissions are intentional; I might omit to do something simply because I do not care, or because I am in a rush, or forget, and so on. But this itself indicates that focusing on omissions just as such is unlikely to be helpful in making moral assessments.

The second part of my defense is to point out that categorization of a non-action as an omission is generated by the existence of prior duties, habits, and expectations; there is no natural class of non-actions that are "omissions." If I have a duty to feed Smith, or if I habitually feed him, or if I led him to believe that I would feed him, then my not feeding will be an omission. But there are, no doubt, many Smiths in hospitals across the country that I am not currently feeding. These non-actions are hardly to be considered omissions. So whether something *is* an omission or not is context-dependent in a way that whether something is intended is not. Omissions do not form a significant ontological category in the way that intentions and intentional actions do. Action and omission simply do not, therefore, offer us a uniformly important ethical distinction in the way intending and foreseeing do (see Tollefsen, 2006).

In consequence, it is just not that important whether withdrawing food, or any other form of care or therapy constitutes an omission or an action. Some mistakenly think it is when they argue that there is a difference between withdrawing the care, and simply not providing it in the first place. What matters, by contrast, is what the agent intends in her actions or omissions: is the intention to achieve some end by way of the death of the patient? If so, then whether the instrumentality that brings about the death is active behavior, or a mere refraining, the action is intentional killing.

## 13.3 ANH is Futile Care

The concept of "futility" is frequently misunderstood in end-of-life debates, including debates about ANH. To get a handle on it, we need to notice first that futility is itself a notion that makes sense only in a context in which we are trying to do something, only in a context in which we have some end in view. In the absence of an end, nothing is futile – mere facts of nature, for example, can hardly be considered futile.

Something is futile, however, if it is strictly of no benefit in the pursuit of the end for which it has been chosen. If I wish to take my temperature, and my thermometer is broken, it is futile to put it in my mouth, futile to read it, etc. However: mistakes about whether something is futile abound when the end is not carefully distinguished. Suppose my aim in putting a thermometer in my child's mouth is to comfort her, for her worries are relieved when it seems I am concerned for her health. If the thermometer is broken, my actions are not thereby rendered futile, for, relative to my end in view, there is some chance of success.

Failure to recognize this generates both of the mistakes listed above. This is clearly seen in the claim that ANH is futile. For the provision of food and water is futile only, in almost all circumstances, when it fails to provide nutrition – for providing nutrition just is the end which we hope to achieve by feeding someone, certainly when we feed someone who is sick. Now, relative to this end, ANH will be futile only when the patient can no longer take in and process nutrition, as happens occasionally, particularly to some non-PVS patients, at the very end of life. I will return to this point in discussing Mistake Eight. But when ANH is called futile, it is typically so called in relation to ends for which nutrition and hydration are not being currently used as means: the ends of full recovery, or of conscious awareness, etc. But even for a patient who will never regain consciousness, ANH is typically very effective – i.e., not futile – in pursuit of its appropriate end: patients in PVS who are given food and water receive nutrition which helps them maintain their organic existence.

## 13.4 It is Permissible to Accept Futile Care

Of course, the notion of futility has a broader application than simply to ANH and PVS cases: for any given patient, dying or not, a wide range of possible actions might be futile, i.e., might fail to provide any of the benefits for which they were being sought. In these other cases, of course, mistakes very similar to the ANH case can be made: some life-saving surgery might be called "futile" because of the quality of life of the patient after the surgery. But this would not really be a genuine case of futility: life-saving surgery saves a life.

However, suppose that relative to the end for which it is being chosen, some proposed intervention really is futile: the surgery does not save life; or, the medicine taken does not prevent stroke; or, the rub-on lotion does not, in fact, re-grow hair. Such interventions are truly futile. What is the moral relevance of this?

First, it should be clear that judgments of futility as such really are the province of the doctor: the doctor often knows, as the patient does not, whether some course of treatment really will offer the hope of some benefit, however, slight. This is different from the claim earlier in this paper that patients are in the best position to judge the relation between the benefits and the burdens of some proposed intervention. Rather, in the case under consideration, there are no benefits to the proposed intervention.

It seems that in light of this, it would be immoral, a failure of the doctor's professional office, for the doctor to propose an intervention she judged futile, and, similarly, it would be immoral, because a waste of resources, time, money and labor, for a patient to accept, or insist upon, receiving a treatment that had been judged futile in the appropriate sense by an expert doctor. Of course, patients may and often should get second opinions. But what could be the point of doing something futile? It is, in consequence, a mistake to think that futile care or treatment "may be accepted."

## 13.5 All Rejection of "Ordinary Care" is Suicidal

To address this error, it is necessary to say first what suicide is; then to articulate the role that the ordinary/extraordinary distinction is supposed to play in guiding end of life decisions.

As Pope John Paul made clear in the encyclical *Veritatis Splendor*, the foundations of morality are to be found in fundamental human goods, goods the pursuit of which are fulfilling for human agents (John Paul II, 1993). Among these fundamental goods, as the Pope also made clear in *Evangelium Vitae* is the good of life (John Paul II, 1995, no. 34). It is not great step from recognition of these basic truths of practical thinking that actions that are destructive of the good of life are to be avoided.

At the same time, however, it is clear that further specification is necessary of what is to count as action destructive of life. For if any action that had as its causal consequence damage to human life were ruled out, then it would seem to be impossible to act in a moral fashion; for even when I take my children to the park, as utilitarians such as Jonathan Glover have pointed out, I thereby do not spend that time working for the life and health of the millions around the world whose lives are in jeopardy (Glover, 1990).

Similarly, if any action that I took that had as a causal consequence my own death was suicide, then it would be impossible to take morphine for pain, or to refuse life-saving medical treatment, without committing suicide.

In consequence, as discussed in the first section of this paper, the Catholic moral tradition has held that what is impermissible are direct, or intended damagings or destroyings of instances of human life. If, in order to end my pain, I take morphine with a view to ending my life, then I intend my death, and so act suicidally. But if I take morphine to ease the pain, recognizing that a perhaps inevitable consequence of this is that I shall die, but in no way choosing my death as either an end or a means, then I do not intend my death but accept it as a side effect.

Now clearly, not all acceptings of side effects are morally permissible simply because they are side effects, and not part of what is intended. The chemical plant owner who, in order to save money, pollutes the nearby river with his waste, despite the fact that downstream residents will be harmed, need not intend that harm. But accepting it as a side effect is grossly unfair. Considerations of fairness are thus crucial in evaluating the side effects that we permit to affect others when we act.

Such considerations can be important for one considering end of life treatment, as can various other considerations. Any such treatment will no doubt offer various benefits (lest it be futile) yet also threaten certain harms. Some treatments save life, yet leave one relatively incapacitated, or permanently in pain, etc. Should one choose to accept the benefits, or to avoid the harms? As mentioned above, whenever possible, this is a decision for the patient herself to make. Yet how is she to make it? The Church speaks to this by asking whether the benefits are proportionate to the harms or not – is this minimal extension of life proportionate to the harm of permanent nausea and incapacitation for the remainder of one's days, for example.

The best recent work on this question of determining what is proportionate and what not is that of Germain Grisez and Joseph Boyle, both of whom argue that an adequate answer, from the standpoint of the agent in question, is one that takes into account the agent's personal vocation and responsibilities (see Boyle, 2002; Grisez, 1993, pp. 519–545). "Proportionate" is thus context sensitive; given one set of commitments and responsibilities, the benefits of prolonging life might be proportionate to the harms, but given a different set, these benefits might not be. To say, then, that a treatment offers benefits that, for this agent, are proportionate to threatened harms seems precisely what is meant by saying that a course of treatment is "ordinary." It is certainly a mistake to gloss "ordinary" by reference to how frequently a procedure is used, or to how many patients choose it. Rather, ordinary treatment just is treatment whose benefits are proportionate to its burdens for the agent in question.

It follows that it is always *unreasonable* for an agent to refuse a treatment whose benefits were proportionate, i.e., an "ordinary treatment." But it does not follow that this refusal must be suicidal. For the threatened burdens are real, and it is intelligible for an agent to seek to avoid them. An agent might refuse treatment precisely because she does not want to spend the money, or because she does not want to be in pain, even when the reasonable thing to do would be to spend the money or accept the pain. In declining the treatment to avoid the harms, the agent does not in any way intend her own death; she has no suicidal will. But like the chemical plant owner, she accepts a side effect, in this case, her death, which it is unreasonable to accept. She may be guilty of cowardice, miserliness, or of some other vice, but she is not guilty of suicide.

## 13.6 No Rejection of "Extraordinary" Means is Suicidal

Mistake six, like seven to follow, mirrors some of the confusions already addressed. When it is said that it is permissible to refuse "extraordinary" treatments, "extraordinary" cannot refer to some descriptive feature, such as rarity of use. Rather, again an assessment of benefits and burdens is necessary on the part of a patient to see whether those benefits are proportionate to the burdens. When, in light of that reflection, it is determined that benefits are not so proportionate, the treatment may be deemed extraordinary.

So, consider a patient who has considered possible treatment for cancer. The treatment is expensive, and has other unpleasant side effects: severe nausea, weakness, and the possibility of death sooner than the cancer. On the other hand, if successful, the treatment could prolong life beyond what the current prognosis indicates. Now consider further that the patient has no significant family responsibilities; her children are grown, her husband dead. She has some savings, which she hopes to give to her children and grandchildren, but which would quickly be drained by the treatment. And her treatment would require that one of her children leave a job and home in another state to come and provide regular care for her.

It might be that all the children in this case are quite willing to do what is necessary to help the woman receive the treatment; this is not a case of caving in to unwilling family members. But it seems also possible that the patient in this case might decide that the prospect of some further life, though certainly good, was not proportionate to the many threatened burdens. Willing only to avoid the burdens, and in no way choosing to end her life, she might judge the treatment extraordinary, and refuse it, accepting her death from cancer as a side effect.

Yet the fact that the treatment could be judged extraordinary, and indeed, the fact that it is judged extraordinary, by the agent, does not itself mean that the agent therefore chooses in a non-suicidal way. For she might choose to refuse the treatment precisely to end her life, thereby, perhaps, to save herself some suffering, and save her children some money. The judgment that the treatment could be reasonably refused as extraordinary in no way causes the agent to refuse it as extraordinary; it simply opens up a rightful avenue of choice for the agent, an avenue she may yet refuse to go down.

## 13.7 It is Permissible to Accept Extraordinary Treatment

Suppose that the agent judges that the proposed course of treatment is extraordinary, i.e., disproportionate in the benefits it offers to the burdens it threatens. This judgment seems to raise a difficulty that is not much noticed in discussion of the distinction. How could it be reasonable for an agent to choose to accept what she had judged was disproportionate, and hence unreasonable? Should it not rather be the case that an agent who had so judged was under an obligation not to choose the suggested course of treatment?

The difficulty is sometimes overlooked because it makes sense to think of the possible courses of action as all, in themselves, morally permissible. We are not talking about suicidal options, but of choosing to accept or reject treatment, neither of which can be ruled out as impermissible as such in advance. Moreover, perhaps there are cases in which it is not clear to the agent that one option rather than the other is more consistent with her vocational commitments; perhaps, indeed, this unclarity is not merely epistemic, but reflects genuine indeterminacy in the options for that agent.

Yet in such a case the agent would not judge the care rejected to be extraordinary, i.e., disproportionate. To say that it is disproportionate is, on the account given, to say something about the treatment that relates it in a very narrowly specified way to this particular agent, with her particular situations and circumstances. But the judgment that the option is extraordinary is a judgment of conscience that seems to render one option now mandatory and the other forbidden. It is always wrong to do what is unreasonable, and the form of "unreasonable" here at issue has to do precisely with whether an option is disproportionate in the relation of its benefits to burdens. One could not rightly, i.e., reasonably, accept what one had judged disproportionate.

This case is thus somewhat similar to that of the acceptance of futile care. In both cases it is sometimes thought that, since the options are not immoral as such, it is therefore permissible to choose what is either futile or extraordinary. But this is to overlook the vast number of cases in which what is permissible as such is impermissible given certain circumstances of the agent: to date, when one is married, to golf, when one has promised not to, to retire early, when one has debts to pay off. To choose what one has judged, in the circumstances, to be futile or extraordinary is likewise to choose wrongly something that might, in some other set of circumstances, be chosen rightly.

## 13.8 It is Never Permissible to Withdraw Artificial Nutrition and Hydration

In his address on the subject Pope John Paul stated that "the administration of water and food, even when provided by artificial means, always represents a natural means of preserving life, not a medical act. Its use, furthermore, should be considered, in principle, ordinary and proportionate, and as such morally obligatory . . ." (John Paul II, 2004).

The description of this procedure as "in principle" ordinary and proportionate is potentially misleading. For it does not seem that the Pope thereby means that it is never morally permissible to withdraw the ANH of any patient in any circumstance. The following considerations bear this out.

First, the Pope is referring to patients in a permanent or persistent vegetative state, not to any patient whatsoever who might be receiving ANH. Second, the Pope qualifies his claim immediately after the quoted phrase by saying ". . . insofar as and until it is seen to have attained its proper finality, which in the present case consists in providing nourishment to the patient and alleviation of his suffering" (John Paul II, 2004). Now some patients who are not in PVS, but who are in the final stages of a terminal disease, are simply not benefited in this way by ANH; that is, the procedure does not provide life-sustaining nourishment at this end stage of their lives. There seems therefore no reason why it should be maintained for such patients.

Second, the case of PVS patients seems different from some other patients in the following respect as well. There seem to be few burdens of ANH as such for the patient in PVS. There are, no doubt, burdens for family members; this will be discussed shortly. And there is the risk of infection. But beyond this risk, for the patient herself, there seems only to be benefit, the benefit of maintenance of life. She is not uncomfortable, or in any other way inconvenienced by the ANH itself.

By contrast, for a conscious patient, ANH might, itself as such, impose burdens as well as benefits. It might be uncomfortable, or confining. As well, as for the PVS patient, it might carry the risk of infection, a risk the patient might be aware of, and view as a considerable burden. Might these be sufficiently serious burdens for a patient that she could reasonably choose to avoid them, by refusing ANH, with the consequence that she lose her life?

I think this is a difficult case, and perhaps one for which no definite answer from the third person perspective is possible. A slightly easier case has, I think, been suggested by Grisez and Boyle (Grisez, 1997, pp. 223–224; Boyle, 1995, pp. 189–199). Provision of ANH does create burdens for others, including the burden of providing care. This is a burden that may, and should, be willingly chosen, as a form of solidarity with the ill, but it is true that it is burdensome and requires sometimes significant commitments. Could a person, with the prospect of ANH in view, whether in PVS or not, choose to forego all care as a matter of charity, sparing her family the burdens of providing care while recognizing that she would lose her life in consequence? This seems to me morally possible.

If so, however, it leads to a general principle, from consideration of analogous cases. Consider the analogous cases first. It is in general possible to refuse care for morally appropriate reasons, reasons which others might in principle not have full access to. Hence the principle of the autonomy of the patient. But some patients will inevitably abuse that autonomy and choose to refuse care not because it was disproportionate and extraordinary, but precisely to end their lives. Barring a very explicit statement that this was the intention of a patient, it will most often be difficult or impossible to tell whether a patient is acting morally or not.

Now if the possibility of morally impermissible action were the paramount consideration for third parties, it might then seem impermissible to aid in an action that might itself be impermissible. Doctors and family members would be morally prohibited from withdrawing or omitting treatment. Yet this consideration is not paramount: it is the possibility of acting morally, of uprightly refusing treatment, and the locus of authority for this judgment in the patient, that seem properly to guide deliberations of third parties, whose duty is, it seems, to respect the expressed wishes of the ill, even when they are no longer capable of re-expressing them.

Indeed, it seems not only permissible to respect these wishes, but, ceteris paribus, obligatory; doctors and family members are not judges over patients, and they violate the autonomy of those patients by being overly paternalistic. When a patient has made explicitly clear, by some advanced directive, that she does not wish to receive this or that form of treatment, it is the doctors' and family members' obligation to refrain from providing it. Thus even for the Jehovah's Witness in need of a blood transfusion, respect for patient autonomy requires that doctors and family members refrain from providing what will obviously save life, and seems to most morally permissible in itself.

It seems, then, that these cases should be considered parallel with patients on ANH, whether in PVS or not. Where there is, or has been, an explicit statement that ANH is not desired, that the patient wishes to refuse it, in the event that it is recommended, then, because this decision can be made uprightly, doctors and family members have a duty to respect the stated wishes of the patient. This view is reflected in the current state of the law; since *Cruzan*, it is clear that an explicit statement of intention to refuse care requires withholding or withdrawing that care, a position distinguished by the Supreme Court in *Vacco v. Quill* from the claim that there is a right to assistance in deliberately ending one's life (U.S. Supreme Court, 1990; U.S. Supreme Court, 1997).

So to the question: is it ever permissible to remove (or not start) a patient's ANH, the answer seems to be "yes," when there is or has been an explicitly stated desire for this on the part of the patient. Even in those cases in which the patient might have so desired for wrongful reasons, there is no need for those who withdraw ANH to formally cooperate in the patient's wrongful choice; their intention can be rather to act so as to respect patient autonomy. This seems consistent with the Pope's claims;

surely, if he thought that following the requirements of the law in these matters was morally impermissible, he would have said so.

## 13.9 The Burdens that Might Permissibly be Avoided Include the Burdens of the Condition Itself

The penultimate mistake on my list is one that it is natural for any sympathetic person to make in light of the radically debilitated condition of the PVS sufferer. Such patients, it must be acknowledged, lead radically disabled lives. They cannot flourish in the manner of agents possessed of full consciousness and the wide range of cognitive abilities of a healthy adult.

For this reason it is tempting to see the burdens that may rightly be avoided through permissible action as including the burdens of the condition itself. If the burdens of financial strain, or painful procedures can be avoided, with death foreseen as a side effect, why not the burdens of the condition itself?

The answer to this must return to the structure of the action aimed at avoiding other burdens. When financial cost is a burdensome side effect of some possible treatment, how is it that the agent seeks to avoid suffering this burden? The answer is that the patient chooses not to receive the treatment in order to avoid having to pay for the treatment. And if not receiving the treatment has as its causal consequence the earlier death of the patient, then this death need not be intended, for it is not part of what is chosen as a means – refusing treatment – or an end –avoiding the financial burdens.

Now there are cases in which an agent may legitimately choose something with a view to avoiding burdens associated with a medical condition itself: the patient may take medicine, for example, or undergo surgery, with a view to eliminating the condition, or at least with a view to alleviating its symptoms. And certainly similar choices can be made on behalf of a patient by a proxy, when it is determined that some procedure will restore the patient to health, yet the patient is incompetent to choose that treatment.

But in the case of a patient in PVS, or in the case of someone considering the prospect of life in PVS, what could be chosen by the patient with a view to avoiding the burdens of the condition itself? There is no cure, and no prospect of alleviating the symptoms. Could it be the case that refusing ANH is itself a means to avoiding or eliminating the burdensome condition? It could not: for refusing to receive nutrition and hydration on its own does not accomplish what is desired: it is not as if the presence of food and water has been causing the condition, so that eliminating food and water will eliminate the burdens of the condition. Rather, removal of ANH is effective in bringing the burdens of the condition to an end only by bringing the patient's life to an end.

If this is so, then the choice to remove ANH in order to eliminate the burdens of the condition itself is a homicidal choice, when made by a proxy, and a suicidal choice, when made by someone contemplating the possibility of PVS. It should be stressed that this choice is different from the previously discussed choice to forego ANH because of the burdens of ANH itself, whether those burdens accrue to the patient or to others. At the same time, however, the discussion provides a context for making one final point.

## 13.10 The Burdens that May Permissibly be Avoided in Removing ANH Include the Burden of Care for the Patient

The discussion of Mistakes Eight and Nine might lead someone to the following train of thought. Suppose that Jones is caring for Smith, who is in PVS. Jones might be struck by the argument over Mistake Eight to the effect that someone contemplating a future condition of PVS might direct that all care be foregone, as an act of charity to others; and he might be struck by the argument over Mistake Nine to the effect that it would be homicidal to refuse ANH to avoid the burdens of the condition of PVS itself. But, thinks Jones, the care for Smith, while Smith is in PVS, creates burdens for me. They are not the burdens of the condition itself, but the burdens of paying for Smith's housing, of attending to her health in other respects, of providing the ANH itself, and of emotional wear and tear. Additionally, Jones might be married to Smith, and might feel burdened by the inability to remarry or seek the comfort of others.

So Jones might come to the conclusion that it would be morally acceptable to remove ANH from Smith in order to avoid all these burdens. Jones would merely be duplicating what Smith might permissibly do in refusing care out of charity for others, and would not be duplicating anything that Smith might do in refusing ANH because of antipathy towards the condition of PVS.

This argument has, it seems, exerted considerable appeal over the years, and its acceptance might explain why removal of ANH became a live option even in some Catholic hospitals. And yet it seems to me that it was decisively refuted by Grisez almost twenty years ago by means of two considerations (Grisez, 1990). The first concerns the burden of ANH itself. This burden is relatively light: once established, nutrition and hydration can be provided by non-medical persons with little oversight, and it can be provided at home, rather than in a hospital, for a moderate cost.

Jones clearly does suffer from genuine burdens, namely, the burdens listed above. But ANH is not responsible for those burdens except insofar as it is responsible for maintaining the patient's life. It is the patient's continued existence in PVS that creates burdens for Jones, and the only way that removal of ANH from Smith alleviates these burdens is by eliminating Smith.

This is not, it should be noted, an argument about the natural causality of removing ANH; if it were, then every instance of its removal, and the removal of other life sustaining technologies, would be morally impermissible, since causally responsible for the patient's death. Rather, the argument reflects the structure of Jones' practical deliberation: Jones thinks, "I wish to be free of these burdens," and suggests to himself the removal of ANH as a means. Why does Jones think that removal will bring about the desired state of affairs? Because removal of ANH will mean the end of Smith, whose continued existence is so burdensome. The end of Smith is thus part of Jones' intention, chosen as the proximate end of removal of ANH, and as a means to the elimination of the burdens Jones suffers.

The second consideration concerns a possible objection based in the discussion of Mistake Eight. There I argued that an agent, say, Smith, might, in advance of PVS or some other condition, choose as an act of charity to spare her friends and relatives the burdens of care for her. But if Smith can so choose, why can Jones not so choose? Grisez's answer is simple and decisive: one cannot exercise charity on behalf of another. Jones may not be charitable on Smith's behalf in this case any more than he can volunteer her money to hurricane relief on her behalf.

In consequence, there is no way, consistent with the prescription against intentional killing, that Jones can remove ANH from Smith, either to reduce her burdens or to reduce his. Her burdens, apart from the burdens of the condition itself, are relatively light, and his burdens, to the extent that they are significant, stem from Smith's existence, rather than the provision to her of ANH.

What is the appropriate response, then, for Jones, faced with what must be acknowledged to be a tragic situation? It must be noted that identifying moral restrictions and obligations regarding Jones' treatment of Smith in no way minimizes the possibly heroic nature of Jones' response; no tragic situation requires easy choices and actions. Still, Jones' situation is not tragic in the sense that there is no possible good to be pursued. Jones can continue to care for Smith as part of a larger effort to maintain solidarity with her. If Jones is married to Smith, this will require fidelity to that marriage.

While this is very hard, there is testimony from those who have pursued precisely such a course that their lives have been enriched, rather than the opposite, by their continuing love and care of a disabled partner or family member. This enrichment might be seen as the enrichment, with obvious differences, that comes from caring for very young children, who are also unable to reciprocate, and who require unconditional and full time care. Of course, such children typically grow to maturity; but if, at the age of three months, a child was stricken with a condition that meant that she would not measurably develop intellectually and socially through the rest of her life, this too would not eliminate the possibility of benefit in providing care for her.

The ten mistakes identified above should not, I think, be seen in themselves as consequences of moral failure; as mentioned early in this paper, these are difficult issues and cases, where confusion and disagreement among people of good will is possible. Nevertheless, it seems obviously to be the case that the more, as a culture, we are divided between those who understand the lives of those in PVS as degrading or meaningless, or the lives of non-persons, and who see the choice to provide care for such patients as pointless and absent of any good, and those who see an imperative merely to fight back against a culture of death, to that extent it will become progressively more difficult for anyone on either side to make clear headed distinctions and reasonable choices in the care of the permanently disabled.

# References

Aquinas, T. (1996). *Summa Theologica.* In A. Ross & P.G. Walsh (Eds.), Blackfriars edition. New York: McGraw-Hill.

Boyle, J. (1989). Sanctity of life and suicide: Tensions and developments within common morality. In B. A. Brody, (Ed.), *Suicide and Euthanasia* (pp. 221–250). The Netherlands: Kluwer Academic Publishers.

Boyle, J. (1995). A case for sometimes feeding patients in persistent vegetative state. In J. Keown (Ed.), *Examining Euthanasia: Legal, Ethical and Clinical Perspectives* (pp. 189–198). Cambridge: Cambridge University Press.

Boyle, J. (2002). Limiting access to health care: A traditional roman catholic analysis. In H. T. Engelhardt & M. Cherry (Eds.), *Allocating Scarce Medical Resources: Roman Catholic Perspectives* (pp. 77–95). Washington, DC: Georgetown University Press.

Brugger, E. C. (2003). *Capital Punishment and Roman Catholic Moral Tradition.* Notre Dame, IN: University of Notre Dame Press.

Glover, J. (1990). *Causing Death and Saving Lives.* New York: Viking.

Grisez, G. (1970). Toward a consistent natural-law ethics of killing. *The American Journal of Jurisprudence, 15,* 64–96.

Grisez, G. (1990). Should nutrition and hydration be provided to permanently comatose and other mentally disabled patients. *Linacre Quarterly, 57,* 30–38.

Grisez, G. (1993). *The Way of the Lord Jesus: Vol. 2: Living a Christian Life.* Quincy IL: Franciscan Press.

Grisez, G. (1997). *Difficult Moral Questions: The Way of the Lord Jesus, Vol. 3.* Quincy, IL: Franciscan Press.

John Paul II. (1993). *Veritatis Splendor.* Vatican City: Libreria Editrice Vaticana.

John Paul II. (1995). *Evangelium Vitae.* Vatican City: Libreria Editrice Vaticana.

John Paul II. (2004). Care for patients in a "Permanent" vegetative state. *Origins, 33,* 737, 739–740.

Pius XII. (1958). The Prolongation of Life. *The Pope Speaks, 4,* 395–398.

Tollefsen, C. (2006). Is a purely first person account of human action defensible. *Ethical Theory and Moral Practice, 9,* 441–460.

U.S. Supreme Court. (1990). Cruzan v. Director, MDH, 497 U.S. 261.

U.S. Supreme Court. (1997). Vacco v. Quill, MDH, 521 U.S. 793.

# Index

# Philosophy and Medicine

1. H. Tristram Engelhardt, Jr. and S.F. Spicker (eds.): *Evaluation and Explanation in the Biomedical Sciences.* 1975      ISBN 90-277-0553-4
2. S.F. Spicker and H. Tristram Engelhardt, Jr. (eds.): *Philosophical Dimensions of the Neuro-Medical Sciences.* 1976      ISBN 90-277-0672-7
3. S.F. Spicker and H. Tristram Engelhardt, Jr. (eds.): *Philosophical Medical Ethics.* Its Nature and Significance. 1977      ISBN 90-277-0772-3
4. H. Tristram Engelhardt, Jr. and S.F. Spicker (eds.): *Mental Health.* Philosophical Perspectives. 1978      ISBN 90-277-0828-2
5. B.A. Brody and H. Tristram Engelhardt, Jr. (eds.): *Mental Illness.* Law and Public Policy. 1980      ISBN 90-277-1057-0
6. H. Tristram Engelhardt, Jr., S.F. Spicker and B. Towers (eds.): *Clinical Judgment.* A Critical Appraisal. 1979      ISBN 90-277-0952-1
7. S.F. Spicker (ed.): *Organism, Medicine, and Metaphysics.* Essays in Honor of Hans Jonas on His 75th Birthday. 1978      ISBN 90-277-0823-1
8. E.E. Shelp (ed.): *Justice and Health Care.* 1981
     ISBN 90-277-1207-7; Pb 90-277-1251-4
9. S.F. Spicker, J.M. Healey, Jr. and H. Tristram Engelhardt, Jr. (eds.): *The Law-Medicine Relation.* A Philosophical Exploration. 1981      ISBN 90-277-1217-4
10. W.B. Bondeson, H. Tristram Engelhardt, Jr., S.F. Spicker and J.M. White, Jr. (eds.): *New Knowledge in the Biomedical Sciences.* Some Moral Implications of Its Acquisition, Possession, and Use. 1982      ISBN 90-277-1319-7
11. E.E. Shelp (ed.): *Beneficence and Health Care.* 1982      ISBN 90-277-1377-4
12. G.J. Agich (ed.): *Responsibility in Health Care.* 1982      ISBN 90-277-1417-7
13. W.B. Bondeson, H. Tristram Engelhardt, Jr., S.F. Spicker and D.H. Winship: *Abortion and the Status of the Fetus.* 2nd printing, 1984      ISBN 90-277-1493-2
14. E.E. Shelp (ed.): *The Clinical Encounter.* The Moral Fabric of the Patient-Physician Relationship. 1983      ISBN 90-277-1593-9
15. L. Kopelman and J.C. Moskop (eds.): *Ethics and Mental Retardation.* 1984
     ISBN 90-277-1630-7
16. L. Nordenfelt and B.I.B. Lindahl (eds.): *Health, Disease, and Causal Explanations in Medicine.* 1984      ISBN 90-277-1660-9
17. E.E. Shelp (ed.): *Virtue and Medicine.* Explorations in the Character of Medicine. 1985      ISBN 90-277-1808-3
18. P. Carrick: *Medical Ethics in Antiquity.* Philosophical Perspectives on Abortion and Euthanasia. 1985      ISBN 90-277-1825-3; Pb 90-277-1915-2
19. J.C. Moskop and L. Kopelman (eds.): *Ethics and Critical Care Medicine.* 1985
     ISBN 90-277-1820-2
20. E.E. Shelp (ed.): *Theology and Bioethics.* Exploring the Foundations and Frontiers. 1985      ISBN 90-277-1857-1
21. G.J. Agich and C.E. Begley (eds.): *The Price of Health.* 1986
     ISBN 90-277-2285-4
22. E.E. Shelp (ed.): *Sexuality and Medicine.* Vol. I: Conceptual Roots. 1987
     ISBN 90-277-2290-0; Pb 90-277-2386-9
23. E.E. Shelp (ed.): *Sexuality and Medicine.* Vol. II: Ethical Viewpoints in Transition. 1987      ISBN 1-55608-013-1; Pb 1-55608-016-6

# Philosophy and Medicine

24. R.C. McMillan, H. Tristram Engelhardt, Jr., and S.F. Spicker (eds.): *Euthanasia and the Newborn*. Conflicts Regarding Saving Lives. 1987
ISBN 90-277-2299-4; Pb 1-55608-039-5
25. S.F. Spicker, S.R. Ingman and I.R. Lawson (eds.): *Ethical Dimensions of Geriatric Care*. Value Conflicts for the 21st Century. 1987          ISBN 1-55608-027-1
26. L. Nordenfelt: *On the Nature of Health*. An Action-Theoretic Approach. 2nd, rev. ed. 1995          ISBN 0-7923-3369-1; Pb 0-7923-3470-1
27. S.F. Spicker, W.B. Bondeson and H. Tristram Engelhardt, Jr. (eds.): *The Contraceptive Ethos*. Reproductive Rights and Responsibilities. 1987          ISBN 1-55608-035-2
28. S.F. Spicker, I. Alon, A. de Vries and H. Tristram Engelhardt, Jr. (eds.): *The Use of Human Beings in Research*. With Special Reference to Clinical Trials. 1988
ISBN 1-55608-043-3
29. N.M.P. King, L.R. Churchill and A.W. Cross (eds.): *The Physician as Captain of the Ship*. A Critical Reappraisal. 1988          ISBN 1-55608-044-1
30. H.-M. Sass and R.U. Massey (eds.): *Health Care Systems*. Moral Conflicts in European and American Public Policy. 1988          ISBN 1-55608-045-X
31. R.M. Zaner (ed.): *Death: Beyond Whole-Brain Criteria*. 1988          ISBN 1-55608-053-0
32. B.A. Brody (ed.): *Moral Theory and Moral Judgments in Medical Ethics*. 1988
ISBN 1-55608-060-3
33. L.M. Kopelman and J.C. Moskop (eds.): *Children and Health Care*. Moral and Social Issues. 1989          ISBN 1-55608-078-6
34. E.D. Pellegrino, J.P. Langan and J. Collins Harvey (eds.): *Catholic Perspectives on Medical Morals*. Foundational Issues. 1989          ISBN 1-55608-083-2
35. B.A. Brody (ed.): *Suicide and Euthanasia*. Historical and Contemporary Themes. 1989          ISBN 0-7923-0106-4
36. H.A.M.J. ten Have, G.K. Kimsma and S.F. Spicker (eds.): *The Growth of Medical Knowledge*. 1990          ISBN 0-7923-0736-4
37. I. Löwy (ed.): *The Polish School of Philosophy of Medicine*. From Tytus Chałubiński (1820–1889) to Ludwik Fleck (1896–1961). 1990          ISBN 0-7923-0958-8

38. T.J. Bole III and W.B. Bondeson: *Rights to Health Care*. 1991          ISBN 0-7923-1137-X
39. M.A.G. Cutter and E.E. Shelp (eds.): *Competency*. A Study of Informal Competency Determinations in Primary Care. 1991          ISBN 0-7923-1304-6
40. J.L. Peset and D. Gracia (eds.): *The Ethics of Diagnosis*. 1992          ISBN 0-7923-1544-8
41. K.W. Wildes, S.J., F. Abel, S.J. and J.C. Harvey (eds.): *Birth, Suffering, and Death*. Catholic Perspectives at the Edges of Life. 1992 [CSiB-1]
ISBN 0-7923-1547-2; Pb 0-7923-2545-1
42. S.K. Toombs: *The Meaning of Illness*. A Phenomenological Account of the Different Perspectives of Physician and Patient. 1992
ISBN 0-7923-1570-7; Pb 0-7923-2443-9
43. D. Leder (ed.): *The Body in Medical Thought and Practice*. 1992
ISBN 0-7923-1657-6
44. C. Delkeskamp-Hayes and M.A.G. Cutter (eds.): *Science, Technology, and the Art of Medicine*. European-American Dialogues. 1993          ISBN 0-7923-1869-2

# Philosophy and Medicine

# Philosophy and Medicine

# Philosophy and Medicine

88. M.C. Rawlinson and S. Lundeen (eds.): *The Voice of Breast Cancer in Medicine and Bioethics*. 2006                                      ISBN 1-4020-4508-5
89. M. Bormuth (ed.): *Life Conduct in Modern Times: Karl Jaspers and Psychoanalysis*. 2006
                                                                    ISBN 1-4020-4764-9
90. H. Kincaid and J. McKitrick (eds.): *Establishing Medical Reality: Essays in the Metaphysics and Epistemology of Biomedical Science*. 2006       ISBN 1-4020-5215-4
91. S.C. Lee (ed.): *The Family, Medical Decision-Making, and Biotechnology: Critical Reflections on Asian Moral Perspectives*. 2007       ISBN 978-1-4020-5219-4
92. H.G. Wright: *Means, Ends and Medical Care*. 2007       ISBN 978-1-4020-5291-0
93. C. Tollefsen (ed.): *Artificial Nutrition and Hydration: The New Catholic Debate*.
                                                                    ISBN 978-1-4020-6206-3

Printed in the United States
141371LV00001B/13/A

9 781402 062063

# DISCARDED
### CONCORDIA UNIV. LIBRARY

CONCORDIA UNIVERSITY LIBRARIES
MONTREAL